RESEARCH METHODS IN PLEISTOCENE GEOMORPHOLOGY

Proceedings:

2nd GUELPH SYMPOSIUM ON GEOMORPHOLOGY, 1971

Edited by:

EIJU YATSU

ALLAN FALCONER

5 . 5 . 78

Published by:
"GEOMORPHOLOGY SYMPOSIUM" Dept. of Geography
University of Guelph, Guelph, Ontario, Canada

in association with
GEO ABSTRACTS LTD., University of East Anglia
Norwich, NOR 88C, England

©Department of Geography, University of Guelph 1972
Geographical Publication No. 2

The cover design is a section in the glacial deposits of N. England.
Photographed by A. Falconer and described in the text (see Falconer
fig. 8)

The volume was printed and bound by
The Alger Press Ltd., Oshawa, Ontario.

Information about the symposium series may be
obtained from "Geomorphology Symposium" Department
of Geography, University of Guelph, Guelph, Ontario,
Canada.

Copies of the published proceedings of the Symposia are
available from the sole distributors;

GEO ABSTRACTS LTD.,
University of East Anglia,
Norwich, NOR 88C,
England

CONTENTS

CHAIRMEN OF SESSIONS

E. Yatsu   Professor, Department of Geography, University of
                    Guelph.

D. C. Ford   Associate Professor and associate chairman,
                    Department of Geography, McMaster University.

J. Terasmae   Professor, Department of Geology, Brock
                    University.

J. Fyles   Chief of the Terrain Science Division, Geological
                    Survey of Canada.

B. D. Fahey   Assistant Professor, Department of Geography,
                    University of Guelph.

R. W. Packer   Professor, Department of Geography, University
                    of Western Ontario.

CONTRIBUTING AUTHORS

J. T. Andrews   Institute of Arctic and Alpine Research and
                    Professor, Department of Geological Sciences,
                    University of Colorado.

P. Beaumont   Lecturer, Department of Geography, University
                    of Durham, England.

M. Church   Assistant Professor, Department of Geography,
                    University of British Columbia.

P. Clément   Professeur, Department de Geographié, Université
                    de Sherbrooke.

A. E. Corté	Departamento Geologia, Universidad Nacional Del Sur, Bahia Blanca, Argentina.
A. Dreimanis	Professor, Department of Geology, University of Western Ontario.
J. M. Dubois	Graduate Student, Départment de Geographié, Université de Sherbrooke.
A. Falconer	Assistant Professor, Department of Geography, University of Guelph.
D. C. Ford	Associate Professor and Associate chairman, Department of Geography, also Associate Member, Department of Geology, McMaster University.
P. Gadbois	Graduate Student, Départment de Geographié, Université de Sherbrooke.
H. Gwyn	Graduate Student, Department of Geology, University of Western Ontario.
L. C. Hodgson	Graduate Student, Department of Geography, University of Guelph.
R. May	Graduate Student, Department of Geology, University of Western Ontario.
W. G. Nickling	Graduate Student, Department of Geography, Carleton University.
J. B. Reynolds	Chief Laboratory Technician, Department of Geography, University of Guelph.
H. P. Schwarz	Associate Professor, Department of Geology, McMaster University.
O. Slaymaker	Associate Professor, Department of Geography, University of British Columbia.
P. G. Sutterlin	Associate Professor, Department of Geology, University of Western Ontario.
P. T. Thompson	Graduate Student, Department of Geology, McMaster University.
U. J. Vagners	Ontario Department of Mines and Northern Affairs, Toronto.
P. J. Williams	Professor, Department of Geography, Carleton University.

PREFACE

It is a pleasure to write some opening remarks for the second volume of the proceedings of the Guelph Symposia on Geomorphology. The Geography Department's geomorphology group has worked hard and well to produce two successful symposia. The response to both has been very encouraging. In spite of the financial pressures which made it necessary to levy a registration fee, there was a substantial increase in attendance. We look forward to a still larger gathering for the Third Symposium, scheduled for 1973, which will focus on research methods in Arctic and Alpine geomorphology.

In the planning of these symposia there are many separate tasks, each of which demands the careful attention of one or more individuals. When the proceedings are to be edited, prepared for publication and printed the tasks are multiplied many times. We cannot hope to thank adequately all involved; those acknowledged below are persons whose contributions can be seen in the permanent record.

Dr. Paul Karrow of the University of Waterloo merits special thanks for the field excursion which he conducted so admirably. The people who went on the excursion expressed their appreciation at its conclusion and we take this opportunity to echo their vote of thanks to Dr. Karrow.

Mrs. P. Rogers and Mrs. L. Cawthra, secretaries of the department of Geography, spent many hours at their typewriters and organised the various mailings. Without their efforts little would have been accomplished.

The painstaking work of the cartographers in the department of Geography is reflected in the high quality of illustrations in this volume. Miss Joan Robertson, Mr. Fred Adams, Mr. D. Irvine and Mrs. S. Towndrow deserve our thanks for their patience and co-operation in the tedious business of compiling the illustrations for publication.

The Department of Information, the Central Duplicating Service, and the Central Reservations and Conference Office of the University of Guelph all assisted in the organisation of the symposium. The work of Barbara Ellesley in Central Duplicating and the Department of Information have been invaluable in preparing the text for the printers. Our task was further facilitated by Mr. B. Reynolds, Chief Technician of the Geography Department, who organised the tape recording of the proceedings and so was able to resolve problems in the reporting of the discussion.

The program committee consisting of Drs. Fahey, Falconer, Thompson and Yatsu devoted a great deal of time and effort to the symposium and it is pleasing to see their efforts being brought to fruition with the publication of this volume.

We are grateful to the geological Survey of Canada for its grant towards the cost of the publication of the symposium proceedings. I wish to record the thanks of the department by acknowledging both the grant from Dr. Fortier, director of the Geological Survey of Canada, and the participation in the symposium of Dr. Fyles, chief of the division of Quaternary Research and Geomorphology and his colleague Dr. W. Shilts.

We are encouraged by the growing number of participants from universities across North America and from provincial and federal government agencies within Canada as well as by the high quality of papers and discussion. We hope that these trends will continue in the forthcoming symposia.

It is with pleasure, then, that I preface the proceedings of the Second Guelph Symposium on Geomorphology and look forward to future volumes in the series.

University of Guelph
May 1972.

Kenneth Kelly,
Acting Chairman,
Department of Geography.

FIELD EXPERIMENTS ON FREEZING AND THAWING AT 3.350 METERS IN THE ROCKY MOUNTAINS OF COLORADO, U.S.A.

Arturo E. Corte
Ambrose O. Poulin

INTRODUCTION

Freeze-thaw action in nature in cold climates is evidenced by breakage and sorting of rocks or loose sediments. We are begining to understand the role of some significant variables affecting this process of sorting (Mackay 1953, Lundkvist 1962, Corte 1963 b, Jahn 1963 and others). Field studies made in Thule Greenland (Corte 1963 a p 78) have indicated that there could be a grain size range in which vertical or horizontal sorting would be expected.

With this idea in mind and also considering that there are no experiments in large scale carried in nature in order to study the effects of many cycles of freezing and thawing of loose sediments, it was planned to perform such outdoor experiment.

In the Rocky Mountains of Colorado there are above 3.300 meters (over the timber line) some large sorted features similar to those of the polar regions (Fig 1-2). Also we find in the Rocky Mountains similar sorted features at the bottom of shallow lakes well below the timber line. Are these features active or are they inactive forms of the past permafrost of the last glacial event?

Regarding sorting by freeze thaw action we can ask several questions: 1) Is this kind of sorting dependent of the grain size of the sediments; i.e. the heterogeneity index 2) Is it dependent on the percentage of the fraction finer than #200 sieve, like the frost heaving susceptibility criterion used for soils in engineering construction? 3) How important is the mineralogical composition of the fines? 4) Can we offer a freeze thaw sorting susceptibility criterion for loose sediments?

With these and other questions in mind this experiment was planned in 1958 carried out in 1959, controlled every year by means of stereophotographs untill 1966. After that date the photo survey will be continued for as many years as possible.

1

Fig. 1 A sorted feature, as it appears at the border of the Green Lakes close to the experimental site.

Fig. 2 A sorted feature developed in the active layer in Thule Greenland.

This paper deals with the set up of the experiment its development during seven years and some conclusions and recommendations.

In a later report it will be informed on the developments after 1971, including some sections of the plots and an analysis with a computer on the surface changes. For this reason this paper should be considered a preliminary one.

This experiment was planned prepared and carried out with the financial support of the USA Snow Ice and Permafrost Research Establishment (SIPRE), later on changed to Cold Regions Research and Engineering Laboratory (CRREL) of Hanover New Hampshire. The total cost of this experiment was about 14.000 dollars.

This experiment could be accomplished thanks to the kind cooperation of Dr. John Marr former Director of the Arctic and Alpine Research Institute of the University of Colorado in Boulder Colorado.

EXPERIMENTAL PROCEDURE

During the autumn of 1959 the senior author of this paper (AEC) visited the Boulder watershed area in order to find a place to carry an outdoor experiment. The purpose of the experiment is two fold:
1)   To find out if sorting by freezing and thawing is related to the percentage of the particles finer than 0.074 (#200 sieve).
2)   To test if there is a grain size in which sorting, lateral or vertical could or could not be expected.

It was observed that at the borders of the Green Lakes, about 3.350 meters a.s.l. near the timber line, there are large sorted features (Fig 1) similar to those found in the active layer over the permafrost (Fig 2), in Thule, Greenland.

It was thought that the Green Lakes region would be the proper place for the set-up of an outdoor experiment. Factors related to the selection of this particular location were:
1)   Its high altitude, with a high freezing index which could produce a freeze thaw layer of 1-2 meters deep (similar to the freeze-thaw layer in Thule Greenland).
2)   It is protected as much as possible from disturbance by intruders, because being the watershed of the city of Boulder it is protected by personnel from the Water Department. As we will see later this was not very successfully accomplished.

3) The location in the vicinity of sorted features similar to the ones observed in Greenland Thule area.
4) Location in the place of loose sedimentary materials: moraine and regolith with different proportions of the fine fraction.
5) It is adjacent to and has access from Science Lodge, a field research facility of the University of Colorado's Institute of Arctic and Alpine Research. In addition to furnishing a significant amount of support (at no cost) the Institute and Science Lodge provide an excellent research environment by virtue of its location and facilities but even more so by the presence of numerous scientists working in ecology, soils frost phenomena, glacial geology etc. Our personal relationships with the colleagues of Science Lodge and in the City of Boulder Water Department were excellent.

The place chosen for the set up of the experiment was a temporary lake fed by a snow bank and surrounded by outcrops of metamorphic rocks.

The process of establishing the site included:
1) Excavation of the in situ soil to bedrock, creating a 30 meters diameter basin with a depth down to 6 meters open to one side.
2) Selection and placement of soils for the individual test plots.
3) Construction of a small dam for the maintenance of a high water table in the soils.
4) Installation of control points for reference of movement observations made from vertical photography.
5) Installation of three thermocouple stations for measuring temperature down to 2.10 deep.

Using a D-8 bulldozer, a large front end loader, 2 trucks, jack hammers, a pump, picks and shovels, this work took two months for six men to accomplish.

DESIGN OF THE EXPERIMENT

Three soil types were selected for the test:
1) A clean sandy gravel, zero percent passing the #200 sieve; this material in the active layer in Thule Greenland did not show horizontal sorting. Such material was placed in layers to a depth of more than 2 meters in plot A (Fig 3). The grain size of this material is shown in Fig 4 and is compared to the ranges of sorting and no sorting found in the active layer in Greenland.
2) A well graded sandy gravel with 7 percent finer than the #200 sieve (0.074 mm); such material in the active layer in Thule, Greenland did show horizontal sorting. Such material was placed in layers to a depth of more than 2 meters in plot B (Fig 3). The grain size of

4

this material is shown in Fig 4 and compared to the ranges of sorting and no sorting found in the active layer in Thule Greenland.

3) A "very dirty" sand gravel with 18 percent passing the #200 sieve; such material also showed horizontal sorting in the active layer in Thule, Greenland. This material was also placed in layers to a depth of more than 2 meters in plot C (Fig. 3). The grain size of this material is shown in Fig 4. Plot D consists of a bottom layer containing 18 percent fines with its upper surface in the form of mounds 20-60 cm high and an upper layer consisting of a coarse gravel providing a cover of approximately 20 cm over the mounds.

As stated earlier a small reservoir was created to provide water for saturating the soil (Fig 3). Since at the begining of the experiment the snow fall did not give enough water to fill up the reservoir, it was necessary to pump water from the Green Lake, 12 meters below and 60 meters away. A gasoline pump and plastic tubing were used. Fig 3 shows the four experimental plots after completion and after the water pumping was completed. Measurements of the surface motions were made photogrammetrically using a specially designed 22.5X22,5 cm format camera and Mylar base aerial film. By means of a bridge-like structure (Fig 5) erected over each soil plot the camera is suitably positioned to obtain photographs of mapping quality. Mapping is subsequently performed by a contractor.

FREEZING AND THAWING PENETRATION

It was planned that soil temperatures should be taken during a whole year. Three eleven-point thermocouple stations were installed in the experiment areas. One station was located in plot A, another in plot B and the third one in plot C. Thermocouples were placed in the following depths:

Thermocouple: No.	1	2	3	4	5	6	7	8	9	10	11
Depth in Cm	0-15	-45	-60	-75	-100	-125	-150	-175	-200	-210	

Temperature was measured once a week from September 19-1959 through September 30, 1960. Also air temperatures were recorded by means of a thermograph.

Freezing line penetration was plotted as a function of time for three stations. Freezing period started at the beginning of April (Fig 6). As it should be expected during the first half of this freezing period in area A with 7 percent fines was freezing faster than area C with 18 percent fines. During the second part of the freezing period (after January) due to the accumulation of a thick snow bank over area B its rate of freezing decreased; as a consequence the rate of freezing of area C was greater. In

Fig. 3 General view of the four experiment sites A, B, C, and D and control points 1, 2, 7, 8, 9 and 10. The site is at an elevation of 3,350 meters a.s.1. in the city of Boulder watershed. Round tubes are thermocouple stations.

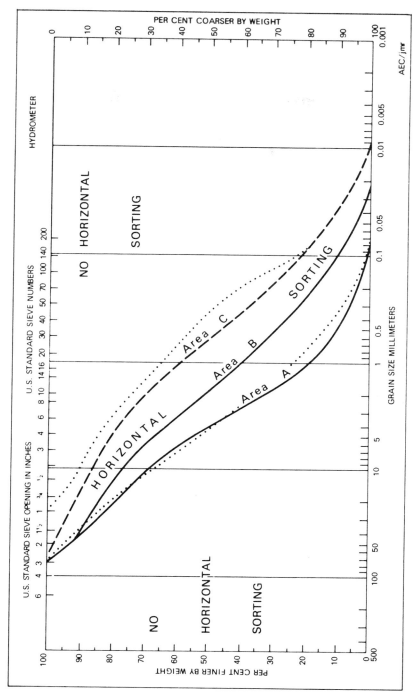

Fig. 4 Ranges of sorting and no sorting found in the active layer in Thule, Greenland, compared with the grain size composition of the materials used in areas A, B, and C. Area D is composed of an upper layer coarser than Material A and a lower layer similar to Material C.

Fig. 5 Scafolding used for taking pictures of the experiment areas.

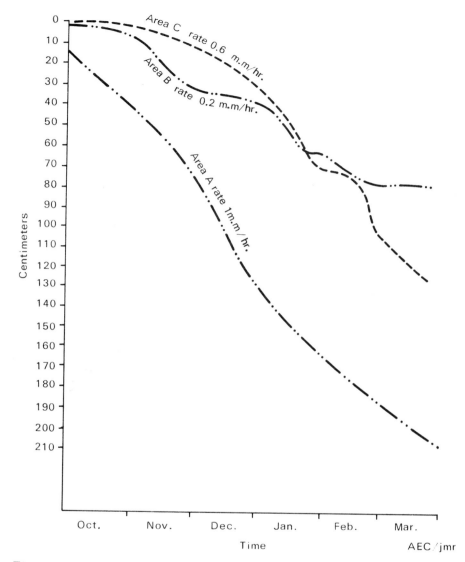

Fig. 6 Freezing line penetration in three experiment areas A-B and C. Area A the coarsest has a deep frost penetration; and fastest rate of freezing. Areas B and C finer areas have a slower rate of freezing, shallower penetration.

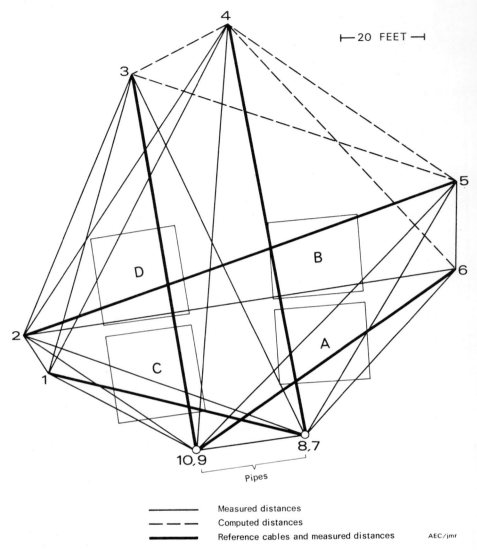

├─ 20 FEET ─┤

Pipes

———————— Measured distances
— — — — Computed distances
━━━━━━ Reference cables and measured distances   AEC/jmr

Fig. 7 Control system for soil movement measurements at the Colorado site. Indicated locations of the soil plots are only approximate.

10

this way the average rate of freezing for area B (7 percent fines) was C,2 mm/hr and for area C (18 percent fines) was 0.6 mm/hr.

These values are in agreement with other values obtained by different authors (Cook 1955, Taylor 1956, Drew and others 1958, Kreutz 1942).

The rate of thawing can not be presented at this moment however it can be stated that thawing took place from the top down and from the bottom up. In a later report a more detailed temperature regime of the experiment areas will be presented.

MEASUREMENT CONTROL

Ten control points were established for measurement reference, six in bedrock, numbers 1-6 and two on each of two, seven foot long pipes embedded in the fill material, numbers 7-10 (Fig 7). The bedrock points are marked on 2 cm steel rods grouted into holes in the rock. The layout of this control system is shown in Fig 7. The system provides horizontal (x,y) reference by virtue of small diameter, stainless steel cables suspended between the control points prior to taking the measurement photography, thereby providing two intersecting lines of known position over each of the soil plots. Vertical (z) reference is provided by assuming one of the bed-rock points as a bench mark from which the elevations of objects in the pictures are referenced. Even though the pipes for points 7-8-9-and 10 are inside larger pipes with the intervening annuluses filled with wax and oil mixture, there was never any doubt that they would be more subject to motion than the bedrock points. However, in some cases the apparent competence of the rock left some question as to the stability of the point placed in it. Consequently, it became necessary to test their stability by a series of annual surveys. Six surveys have been made: 1960, 1961, 1962, 1963, 1964 and 1966. An early snow fall prevented a survey from being made in 1959. The preliminary results from a combined analysis of these surveys seems to indicate that all of the bedrock points may be stable. Assuming that the bedrock points are stable and that, therefore, all of the variations in the control points surveys can be attributed to the movement of points 7, 8, 9 and 10, the next step will be to stablish a grid system in which to reference the movement measurements and to compute:
1)   The fix grid coordinates of points 1-6, and
2)   The annual coordinates of points 7-10 and the cable intersections. It is intended that the grid system will be established with the (x) axis parallel to line 2-5 and with the (y) axis passing through point 2. line 2-5 will be given an arbitrary (y) value of 18 meters, therby resulting in a positive value for all the X and Y coordinates concerned with the survey and with the soil movement measurements.

MAPPING

Except for location B, 1964, maps have been compiled for all the locations for the years 1960, 1961, 1962, 1964 and 1966. However, discrepancies occurred in the 1960 and 1961 maps which we could not believe were surveying errors but yet could be firmly identify as mapping errors and they seemed too significant to pass off. On the supposition that the discrepancies may have resulted from using the wrong focal length in the photogrammetric plotter, it was designed a test of the supposedly 1 to 1 vertical to horizontal scale ratio.

This test consisted merely of the inclusion of a carefully constructed "step model" with known hight difference between each of its four steps in the photography obtained in 1962 and 1963. After analysing the mapped elevations of the steps, it was concluded that the optical model used in mapping had been flattened by approximately 4 percent. This conclusion was subsequently substantiated by the mapping contractor. They had used 153.5 mm instead of the proper 159.8 mm. Therefore, they will remap the areas at their expense.

With the prospect of remapping in mind, it was decided to reconsider the form in which the map data might be presented. There were two reasons for this reconsideration:
1)   It had become obvious that plotting contours at the desired intervals of 6 mm was not practical for location D, were the surface consists mostly of 10 to 15 cm angular cobbles; and
2)   It seemed desirable to present the map in a form more expeditious for analysis.

After talking the problem over with the mapping contractor it was decided that the new maps will consist of a grid system with an interval of 15.2 cm at ground scale. The grid for each soil plot will be referred to the control point survey grid by the computed grid coordinates of the appropriate reference cable intersection. The mapping data will then consist of the x, y and z coordinates at each grid intersection presented in two forms:
1)   Punch cards
2)   Tabular type-out.

In addition to the grid data, the planimetric position of selected stones will be plotted on the maps along with the elevation of the stones high point. A second set of punch cards will contain the elevations of these stones. Presentation of the map data in digital form will greatly facilitate its analysis. But before the remapping can be done it will be necessary to:
1)   Finish the detailed analysis of the annual survey in order to decide whether or not the bedrock control points are stable.
2)   Compute the best values of the distances between the bedrock control points.

3) Compute the fixed coordinates of the bedrock control points and the annual coordinates of points 7, 8, 9, 10, A, B, C, and D (letters are the cable intersections).
4) Compute the changes in reference cable orientation for locations A and C (locations B and D present no problems in this regard because cable 2-5 is fixed).

EXPERIMENTAL RESULTS

At the present stage of this report only qualitative information will be presented. A final version will contain a more precise analysis of the surface changes.

A stereoscopic analysis of the photographs taken during 1960-61-62-63-64 and 1966 shows that:
1) Area A without fines did not develop horizontal sorting. It can be observed that the surface in 1966 is coarser than in 1960. (Figs. 8a - 8b).
2) Areas B and C with 7 and 18 percent fines showed circular depressions in 1960 and 1962, produced by animal tracks or by collapse. Sorting into these depression occurred the year after and then disappeared. (Fig. 9a - 9b). The surfaces of both areas B and C are much coarser in 1966 than in 1960 (Figs. 10a - 10b), (11a - 11b).
3) Area D a cover of coarse gravel with no fines over a layer of well graded sandy-gravel with 18 percent of particles finer than 0.074 mm in the form of mounds, shows the surface coarser in 1966 than in 1960.
4) The development of depressions is an important basic research subject. Other researchers have reported in field studies the formation of depressions (Mackay 1953, Corte 1955, Lundqvist 1962). It should be important to demonstrate if in nature such circular depressions are produced by the freezing process or by animals.
5) Plants, graminiae became established in the coarser plots with no fines areas A and D. In the areas with 7 and 18 percent of particles finer Than 0.074, areas B and C, no graminiae was observed during the seven years of the experiment. (Figs. 8a - b; 10a - b; 11a - b; 12a -b).

This is very probably due to the frost heaving at the root level killing the vegetation. The interesting fact is that such a small amount of 7 percent fines was enough for inhibiting the growth of plants, during the observed years 1960 - 1966.

Table 1 gives the main qualitative results of the experiment.

Fig. 8 *a.* Surface views of plot A in 1960.

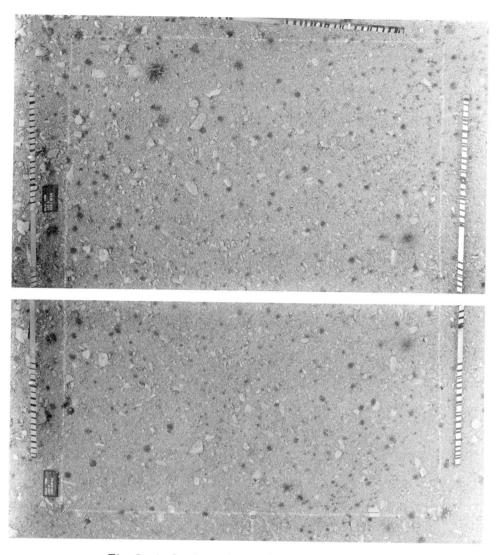

Fig. 8 *b.* Surface views of plot A in 1966.

Fig. 9 *a.* Surface view of plot B in 1962 showing development of depressions.

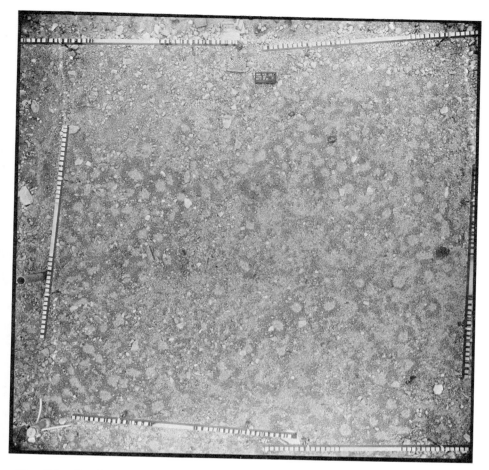

Fig. 9 b. Surface view of plot B in 1963 showing sorting into depressions.

Fig. 10 *a.* Surface views of plot B in 1960.

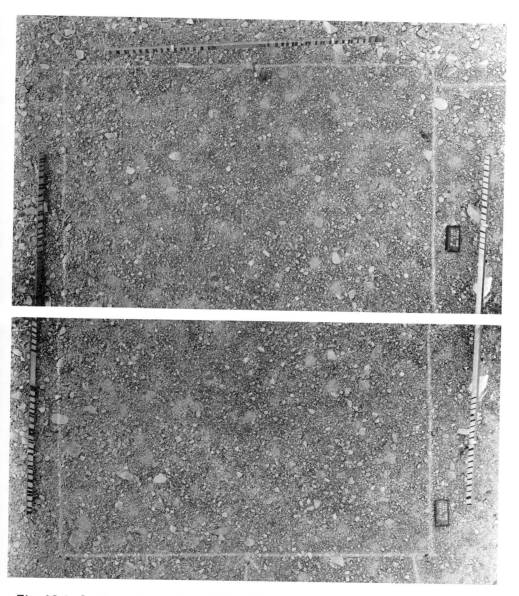

Fig. 10 *b.* Surface views of plot B in 1966 showing sorting into depressions.

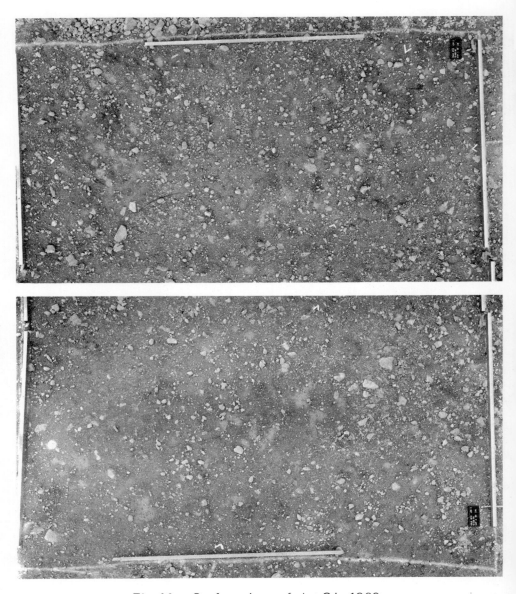

Fig. 11 *a.* Surface views of plot C in 1960.

Fig. 11 *b.* Surface views of plot C in 1966.

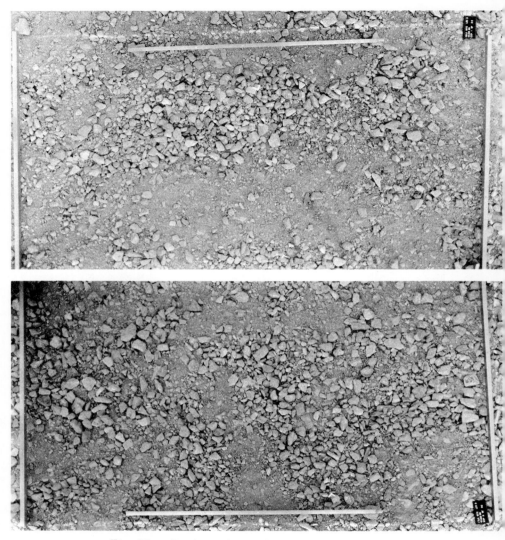

Fig. 12 *a.* Surface views of plot D in 1960.

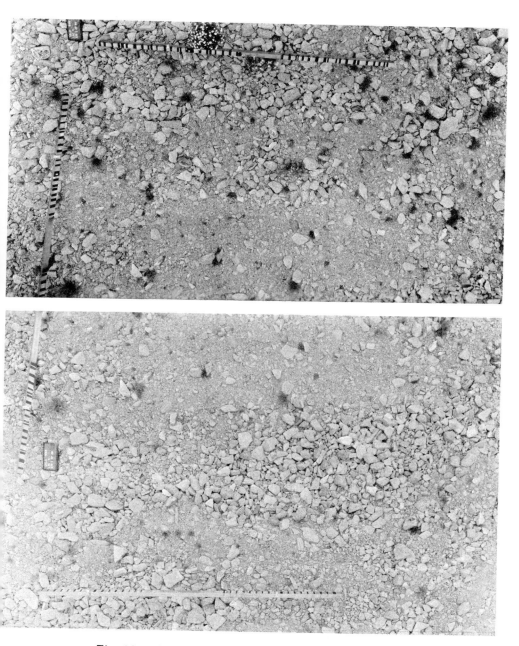

Fig. 12 *b.* Surface views of plot D in 1966.

Areas	A	B	C	D
Exp. started	1959	1959	1959	1959
Percent of particles finer than #200 sieve	0	7	18	Coarse gravel covering material with 18 percent fines
Sorting at the surface	surface in 1966 coarser than in 1960	Sorting into depressions produced by animal tracks in 1961 and 1963.	Sorting into depressions produced by animal tracks in 1961 and 1963.	Surface in 1966 coarser than in 1960.
Vegetation	Surface mottled with graminiae growth	No vegetation.	No vegetation	Graminiae not as dense as in A
Stereo-available from years:	1960-66 except 1965	1960-1966 except 1965	1960-66 except 1965	1960-66 except 1965
General view of the surface after seven years	Surface in 1966 is coarser than in 1959.	Surface in 1966 is much coarser than in 1960.	Surface in 1966 is much coarser than in 1960.	Surface do not show much concentration of coarse particles as in B and C.

Table 1

Qualitative results of a field experiment on sorting at 3,350 meters a.s.l. in the Rocky Mountains of Colorado Boulder USA.

RECOMMENDATIONS

It is recommended to experiment with:
1) Materials having more than 20 percent of particles finer than the #200 sieve.
2) The mineralogical composition of the fine fraction in order to find out its role on sorting.
3) The significant variables in the process of formation of circular depressions.

REFERENCES

Corte, A. E., 1955. Contribucion a la morfologia periglacial especialmente criopedologica de la Republica Argentina: Societas Geographica Fenniae Acta No. 14, 83-102.

Corte, A. E., 1963 a. Relationship between four ground patterns structure of the active layer and type and distribution of ice in the permafrost: Biuletyn Peryglacjalny No. 12, 90 p. Lods Poland.

Corte, A. E., 1963 b. Experiments on sorting processes and the origin of pattern ground: Washington, Natl. Acad. Sci., Natl. Research Council, Proc. Permafrost Internat. Conf., Pub. No. 1287, p. 130-133.

Corte, A. E. and Higashi, A. 1971. Growth and development of perturbations on the soil surface due to repetitions of freezing and thawing: paper presented at the International Geomorphological Conference Caen France, 15 p.

Cook, F. A. 1955. Near surface soil temperatures measurements at Resolute Bay Northwest Territories Arctic V. 8, No. 4, p. 237-249.

Drew, J. V., Tedrow, J.C.F., Shanks, R. E., and Koranda J. J., 1958. Rate and depth of thaw in arctic soils: Am Geophys Union Trans. V. 39 No. 4, p. 697-701.

Jahn, A., 1963. Origin and development of pattern ground in Spitsbergen: Washington, Natl. Acad. Sci., Natl. Research Council, Proc. Permafrost Internat. Conf. Pub. No. 1287 p. 140-145.

Kreutz, W., 1942. Das Eindringen des Frostes im Boden unter Gleichen und verschidenen Witterungsbedingungen Warend des sehr Kalten Winters 1939-40: Reichsamt fur Wetterdienst (Luftwaffe): Wiss., Abh, V. 9 No. 2, p. 1-22.

Lundqvist, J., 1962. Patterned ground and related frost phenomena in Sweden, Sveriges Geologiska Undersokningen Ser C No. 583: 101 p.

Mackay Ross J., 1953. Fissures and mud circles in Cornwallis Island. N.W. Territories: The Canadian Geographer No. 3, p. 31-37.

Schmertmann, J., and Taylor, R.S. 1965. Quantitative data from a pattern ground site over permafrost: USA CRREL Research Rep. No. 96, 76p.

GROUND THERMAL REGIME IN COLD REGIONS

P.J. Williams and W.G. Nickling

Fundamental Relationships

Climate - mean air temperatures

The most widely measured and utilized parameter to characterise climate is air temperature, commonly expressed in the form of monthly and yearly mean values. Mean monthly air temperatures and mean annual temperatures are shown in figure 1 for three places. The air temperature is the result of a very complex set of processes, essentially astronomical, but when considered on a local basis dependent on the magnitude of the components of the energy exchange in the boundary region earth-atmosphere. At any given time and place, air temperatures are affected by the vagaries of weather and specific surface conditions, but are nevertheless usefully represented by mean values, which in turn can with some approximation be represented by a sinusoidal wave as shown in fig. 1.

Ground temperatures

Measurements of the near-surface temperature of the ground on a more or less continuous basis, let us say at 10 centimeters depth, prepared as monthly means give a similar pattern to figure 1. In general the amplitude of the wave (maximum-minimum) is somewhat less than for the air temperature of the place, because of the barrier, or damping, due to the upper 10 centimeters of soil, surface cover etc.

Ground thermal regime is characterised by temperatures expressed as a function of depth and time. If the ground is regarded as a solid homogeneous body, where heat flow occurs only by conduction, it is possible to apply quite simple equations (see for example, Grober, Erk and Grigull, 1961 pp. 81-84) to give these temperatures. The assumption of exclusively conductive heat flow is a gross one, but nevertheless is has been shown by several authors (Pearce and Gold, 1959, Penrod, Walton and Terrell, 1958, Rambaut, 1915) to give a rational and most useful basis for the description of ground thermal regime. This is in sharp contrast to the above-ground case, where air temperature gradients above the ground, or variations with time, can by no means be predicted in this way.

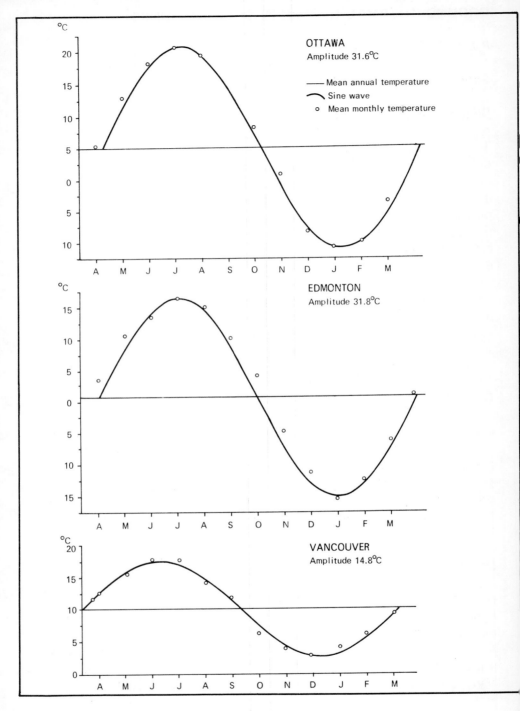

Fig. 1 Mean air temperatures

Basic equations for characterising ground thermal regime

If there is a variation of temperatures at, or, for example 10 cm below the surface of a solid body, which can be represented by a sinusoidal wave of the type in figure 1, then the amplitude of the wave at points deeper is given by:

$$S_z = S_{10} e^{-z\sqrt{\frac{\Pi}{aT}}} \quad \ldots(1)$$

S_z = amplitude at depth z

S_{10} = amplitude at depth 10 cm.

T = period of wave

a = diffusivity $\dfrac{\text{thermal conductivity}}{\text{volume heat capacity}}$ $\quad\ldots(1)$

This is illustrated graphically in figure 2.

Because the conduction of heat is time dependent, the onset of a temperature corresponding to a certain point on the wave (for example the maximum) occurs later with increasing depth. Temperature distributions for two given times are shown in Fig. 3. On August 1st, the ground at 6 to 7 m. depth is experiencing its lowest winter temperature. Computation of such isochrones is somewhat more complex but a simple equation gives the delay $t_2 - t_1$ between the onset of a point on the wave at two depths Z_2 and Z_1 :

$$t_2 - t_1 = \frac{Z_1 - Z_2}{2} \sqrt{\frac{T}{a\Pi}} \quad \ldots(2)$$

A further equation:

$$Z_{\iota} = 2\sqrt{\Pi a T} \quad \ldots(3)$$

gives the depth Z_{ι} at which the delay is a whole period. This depth is often regarded as the depth to which a periodic variation extends. In fact the two curves in fig. 2 (maximum and minimum) approach asymptotically: at this depth the amplitude is reduced to about 1/500th of that at the surface.

A study of these relations leads us to certain conclusions:

1. If the temperature wave is defined for one level it is not necessary to take into account what is actually happening at any level above this, for the purpose of determining temperatures at a lower depth.
2. The amplitude of the wave in the ground is determined by the amplitude at the surface (compare fig. 2). In the examples shown, the amplitude at 10 cm depth (used for calculation) is assumed uniformly 8° less than that of the air temperature. The amount varies with surface cover, etc.
3. The thermal properties of conductivity, and volumetric heat capacity of the soil will affect the amplitude at any depth. In

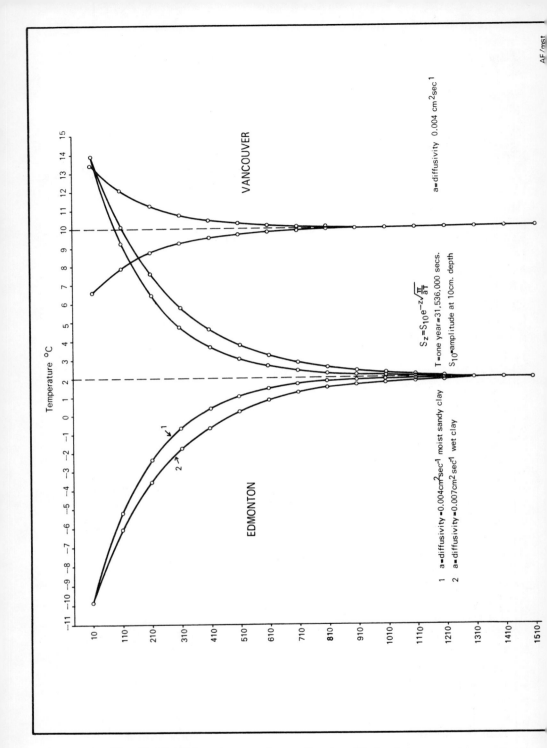

Fig. 2 Calculated annual amplitudes (S_Z) of temperature.

$$S_z = S_{10}e^{-z\sqrt{\frac{\pi}{aT}}}$$

T = one year = 31,536,000 secs.
S_{10} = amplitude at 10cm. depth

a = diffusivity 0.004 cm^{2}sec^{-1}

1 a = diffusivity = 0.004 cm^2 sec^{-1} moist sandy clay
2 a = diffusivity = 0.007 cm^2 sec^{-1} wet clay

VANCOUVER

EDMONTON

Temperature °C

AF/mst

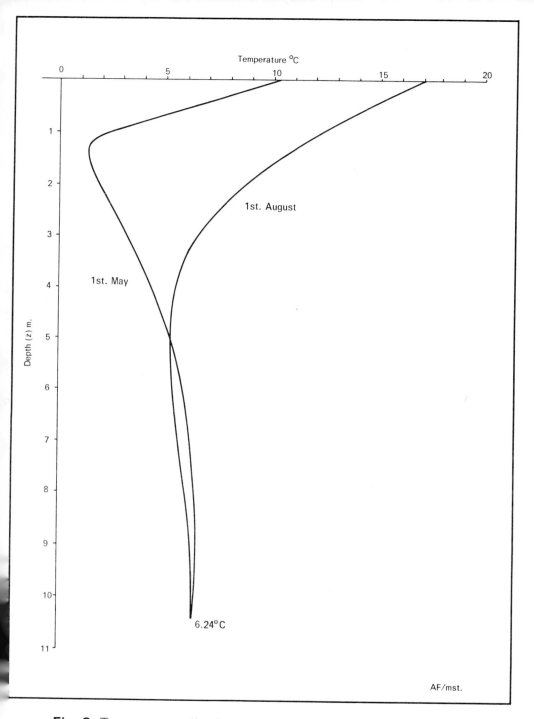

Fig. 3 Temperature distribution with depth, for two occasions
(Oslo, Norway, modified from Klove, 1971).

31

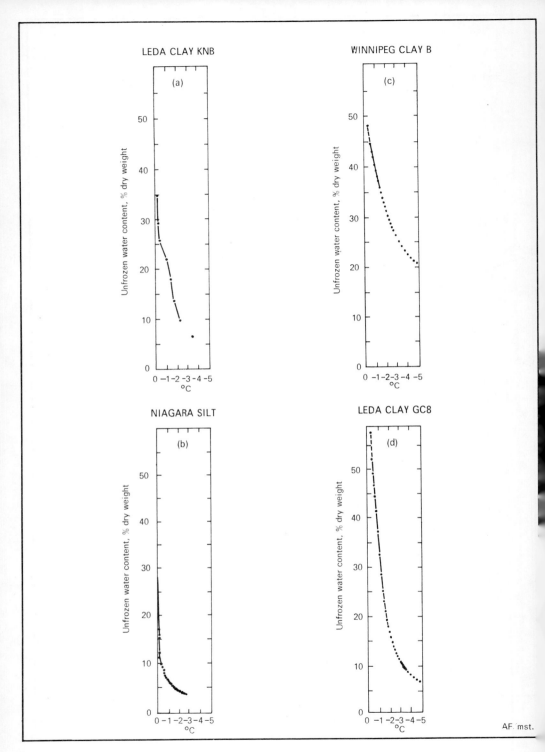

Fig. 4 Unfrozen water content as a function of temperature (from Williams 1964B).

figure 2 values are shown for a wet clay soil as well as a moist sandy clay.

4. The depth to which any temperature wave, originating at the surface, is felt is proportional to the square root of the time (period) for completion of the wave (eq. 3) and of the diffusivity. Thus the diurnal temperature wave occurs to only about $\sqrt{365}$ times (i.e., about 1/19th) the depth for the annual wave at the same location.

5. Equation 3, which contains no amplitude term, gives the depth Z_i at which the amplitude is reduced to a small fraction of its value at the surface. It is often more useful to consider the depth at which the smallest measurable amplitude occurs. This might be e.g., $1/10^\circ C$, and fig. 2 shows clearly that this depth is quite dependent on the amplitude at the surface. The measurable penetration of the diurnal wave is on this account somewhat smaller than suggested under 4, above.

Mean ground temperature

The straight dashed lines in fig. 2 represent the mean annual ground temperature, and its continuation below the depth of annual variation, represents the mean ground temperature. The line should actually have a slight slope, the geothermal gradient, due to a small continuous flow of heat from the earth's interior, and the conductivity of the soil material. The mean ground temperature for a place is fairly similar to the mean annual air temperature. This is not surprising, because a large temperature discontinuity at the surface of the earth would involve a long-term heat exchange whereby the mean ground temperature would be continuously increasing or decreasing, ultimately with catastrophic consequences. On the other hand, since both parameters are the product of complex energy and mass exchanges at the surface of the earth they are not likely to be exactly equal.

This completes the review of ground thermal regime from the point of view of a solid body subjected to a cyclical temperature variation at its surface. The varying and complex nature of the ground surface layer and of the ground itself give a host of deviations from this ideal picture. These are of great importance in the explanation of many observed ground thermal phenomena, especially where freezing and thawing are involved.

Thermal Properties of Freezing and Thawing Soils

Heat transfer in soils

Contrary to the assumptions made for equations (1), (2) and (3) a variety of heat transfer mechanisms in addition to conduction occur. The most important is transfer of heat in association with transfer of mass, whether liquid or vapour. This is particularly obvious following a rain storm when the downward percolating water transfers heat to or removes it from near surface soil layers, producing a substantial deviation from the circumstances described in equations (1) and (2). Temperature effects associated with heat of evaporation or condensation occur frequently in near-surface soils in association with migration of water vapour. During freezing of frost-heaving soils not only does the water migrating to the frost line tend to give its own temperature to the soil but the formation of ice involves the transfer to the soil of a large quantity of heat associated with heat of fusion. Because of the interrelationship of heat and moisture flow in soils, detailed quantitative analysis requires the thermal and moisture regimes be considered together but analytical methods are not yet fully developed. For less detailed studies, the occurrence of various heat transfer processes, on a scale large enough to cause significant deviations from the ideal situations described initially, can often be predicted and allowed for by a knowledgeable observer.

Volumetric heat capacity of freezing soils

Soil is said to freeze when ice is formed in it. However the porous nature of soils influences the behaviour of the soil water, much of which only freezes at temperatures below the normal freezing point. In figure 4 the unfrozen water contents of several soils are shown as a function of temperature. So long as the soil remains at a given temperature (and under unchanged pressure conditions) the proportions of ice and water remain constant. But for any temperature change there is a change in these proportions as indicated by the graphs. Accordingly for the range of temperatures shown any change of temperature must involve the addition or removal of heat of fusion. Because the heat of fusion of water is high (80 cal/gm) this quantity is usually greater than the amount of heat that must be added or removed to change the temperature of the water, ice and mineral soil constituents (their heat capacity sensu stricto). In figure 5 apparent volumetric heat capacities (which include heat of fusion) are shown for the soils: the heat capacities are many times greater than those of the unfrozen soils and are temperature dependent. The values are to some extent dependent on whether the sample has reached the temperature by cooling or warming. In addition when the frost penetrates the ground there will often be a migration of water with attendant additional heat of fusion involved as noted above.

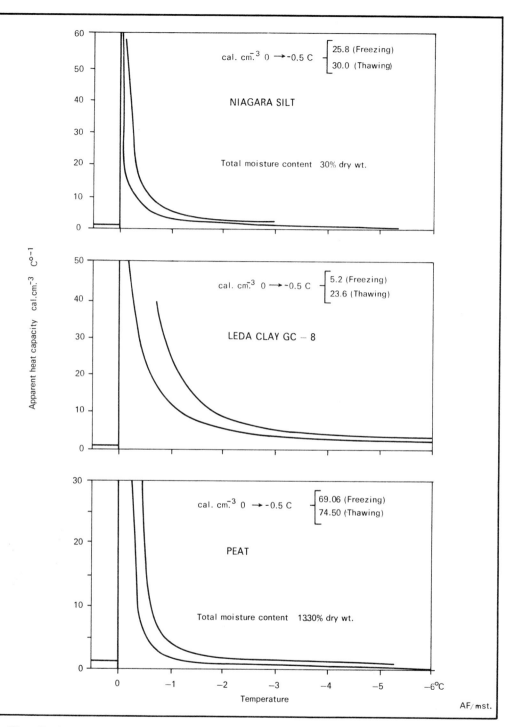

Fig. 5 Apparent heat capacities. The lower curve in each case is the value when the soil is warming, the upper, when cooling. The heat capacity for the interval 0°C to 0.5°C is shown as a single figure (modified from Williams, 1964A).

Thermal Conductivity

The thermal conductivity of ice is 0.005 cal. cm. $^{-1}$sec. $^{-1}$ $^{o}C^{-1}$ while that of water is 0.0013. Consequently there is also a change in thermal conductivity as the temperature of the freezing soil changes. Studies by Penner (1962, 1970) showed the conductivity of a frozen Leda clay to be about .004 cal. cm. $^{-1}$ sec. $^{-1}$ $^{o}C^{-1}$ at temperatures lower than $4^{o}C$, but decreasing to about .002 near $0^{o}C$. A similar (but not identical) soil had a conductivity of about .002, depending on bulk density, in the unfrozen state. A sand with a conductivity of .0006 and wet sandy clay .003 (figures quoted by Beskow 1935) indicate the range of values for unfrozen soils. It appears therefore that the variation of conductivity with temperature in freezing soil will for some practical considerations be of relatively little significance. This is in contrast to the apparent heat capacity, the temperature dependence of which is mainly responsible for the abnormal diffusivities of freezing soil.

Natural Ground Thermal Regime in Cold Regions

Seasonal Frost Penetration

Because the diffusivity of frozen ground is lower than that of unfrozen ground, the pattern shown in figure 2 is modified as shown, for example, in figure 6, and the mean annual ground temperature may also vary with depth. In general the low diffusivity of frozen ground, means that it is less susceptible to temperature change, and characteristically, the frostline is an isotherm which moves very slowly; when freezing or thawing occurs over much of the year, the penetration of the annual wave is much reduced.

Depth of frost penetration depends on the factors of climate, surface, and soil:

1. The effects of *climate* on frost penetration may be considered in terms of different places, which will be assumed to have entirely similar surface and subsurface conditions. Following from the approximate correspondence of air temperatures and ground temperatures, frost penetration will occur when part of the annual wave (figure 1) lies below $0^{o}C$. The area enclosed by the curve below this temperature represents the frost index (^{o}C. days), but this is not by itself sufficient to define the frost penetration. During the winter and spring there is an upward heat flow towards the frozen layer from below. This is still occurring at 1.5m and deeper, on May 31st, in the example,

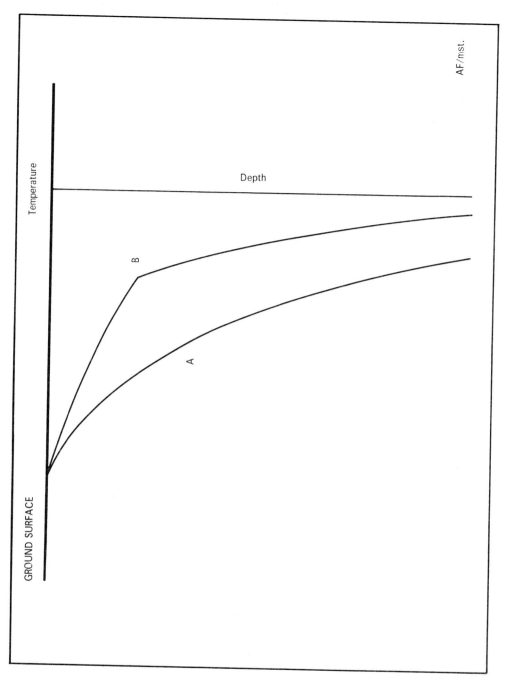

Fig. 6 Effect of latent heat of fusion on annual minimum temperatures. Calculated values for same soil (a) disregarding latent heat, (b) including latent heat effect (modified from Janson, 1964).

figure 3. During the winter it is augmented by heat which entered the ground during the previous summer. In regions with warm summers this heat substantially limits the depth of frost penetration. Such is markedly the case in e.g., Ottawa and Calgary, where the frost penetration is generally less than in the maritime climate of Oslo, despite the occurrence of a higher frost index. Accurate prediction of maximum frost penetration therefore requires knowledge of the upward heat flow to the frost line. At present such information is not widely available, and a qualitative assessment is best made using the mean annual air temperature in combination with frost index values (Williams, 1961).

2. The depth of frost penetration is highly dependent on the *thermal properties of the soil materials.* The least depth occurs (other conditions assumed constant) where the volumetric ice content is great. The latter is the case for soils exhibiting substantial frost heave, (i.e. an accumulation of ice) and also for soils, notably peat, containing much water which freezes at temperatures near $0^{\circ}C$. In soils with a low ice content, variations in thermal conductivity become important. Soils of very low moisture content commonly show relatively small frost penetration, which is ascribable to the low conductivity of the air present in many of the pores. W.G. Brown (1964) shows that, other factors being equal, variations in thermal properties are responsible for a range of variations of up to 2 to 1 in frost penetration.

3. Variations of *surface conditions* drastically affect frost penetration. An extreme example is the presence or absence of a snow cover, the low conductivity of which even in subarctic areas may totally prevent any seasonal frost penetration.

In summary local variations in surface or soil conditions commonly produce variations in frost penetration of a magnitude comparable to those caused by differences in air temperatures associated with locations 100 miles or more apart. The depth of the active layer (i.e., the annual frost penetration) in permafrost regions is related to various factors in a similar manner.

Distribution of Permafrost

In the most northerly regions ground which is frozen throughout the year underlies most of the land surface ('continuous permafrost'). Southwards the permafrost underlies only parts of the ground surface ('discontinuous permafrost') and occurs as a (horizontal) network, or closely or more widely spaced bodies (tens to thousands of feet apart), the bodies becoming more isolated with milder climate. Permafrost occurs below the depth affected by the annual variation, where the mean ground temperature is less than freezing point. It may also extend upwards into the area affected by the annual variation of temperature where this does not involve temperatures above freezing point.

As a starting point in explaining distribution of permafrost, we may consider the mean annual air temperature isotherms, but because of the difference:

mean annual air temperature - mean ground temperature = $\Delta\Theta$

such isotherms will not accurately describe its distribution. The mean annual air temperature on a local basis does not show the distribution that permafrost bodies have in the discontinuous region. In most cases the mean ground temperature is higher by 1 to 4°C and sometimes more, than the air temperature. The magnitude of $\Delta\Theta$ can only be assessed by a careful examination of the microclimatic (primarily surface) conditions at a location. Often a heat exchange process or processes operative during only part of the year is primarily responsible. For example the existence of a thick snow cover in the winter because of its low conductivity, effectively dampens the winter part of the ground temperature wave (Gold 1963). The result is to produce a higher mean annual temperature than in the absence of snow (greater $\Delta\Theta$). An extreme example, is the effect of a sufficiently deep lake which because of the thermal overturn, has water at or near 4° C lying at its bottom for much of the year. The heat capacity of the water and the absorption of radiation at its surface are also relatively more significant than the loss of heat due to evaporation. Even in continuous permafrost zones, a hole exists through the permafrost beneath such lakes. A high rate of evapotranspiration, such as occurs in peat bogs in the summer may give a relatively lower mean ground temperature than elsewhere locally (Williams 1968). Many detailed studies of ground surface conditions in relation to permafrost distribution have been carried out (Brown 1966, 1970, Gold, 1967). It is virtually impossible to describe all the energy exchange processes occuring (to produce a complete energy balance equation), so that correct interpretation of the factor or factors responsible for the permafrost distribution observed must rely in large measure on the general knowledge of ground surface conditions and heat exchange processes, and of the principles outlined, on the part of the observer.

The depth to which permafrost occurs can be related to the mean ground temperature at the upper surface of the permafrost and the geothermal gradient. In figure 7 the intersection of the mean ground temperature against depth line with the freezing temperature ($\simeq 0^\circ$ C) represents the lower extent of the permafrost. However, this procedure assumes a steady state condition, that is, that there are not temporal variations in climate, surface cover or ground conditions. Climate of course changes, and in recent years climatic amelioration has often resulted in a demonstrable rise of the mean ground temperature. Such effects are felt at increasing depth with time. Clear evidence is given by the occurrence of permafrost bodies first at a greater depth than is reached by the annual frost penetration. In the absence of other changes, the existence of the permanently unfrozen layer above the permafrost indicates a progressive warming from the surface. The body of permafrost must then be regarded as relict and only existing because of the time required for the thawing of the mass of ground under the heat flow conditions established. In a similar manner, the onset of permafrost conditions at the surface results ultimately in a thickness of permafrost dependent on the (near surface) mean ground temperature; prior to this the thickness will be proportional to the square root of the time from the onset (compare equation 3).

The Problem Reviewed

Ground thermal regime has been considered from three aspects: the basic laws of heat flow, as they apply to a solid body subjected to cyclic temperature variations at its surface; the values to be ascribed to the basic thermal properties of freezing soils; and the complications introduced by nature in providing an infine variety of conditions, and combination of conditions, of ground surface, soil, and climate. Only a few examples have been given of the numerous circumstances which may be involved in the description of the ground thermal regime at any specific site.

Ground thermal regime will never be understood by a purely theoretical approach, since all the complications that nature can provide will never be foreseen. Neither will the ground thermal regime be understood solely by field observations. Without an understanding of the basic scientific laws, it would be quite impossible to arrange the infinite number of field observations that could be made, in any meaningful and useful framework.

This paper is concerned with fundamentals, rather than easy methods for quantitative prediction. There are methods used by engineers for prediction for example of depth of active layer, but in their simplicity some of the most useful lead the unwary astray.

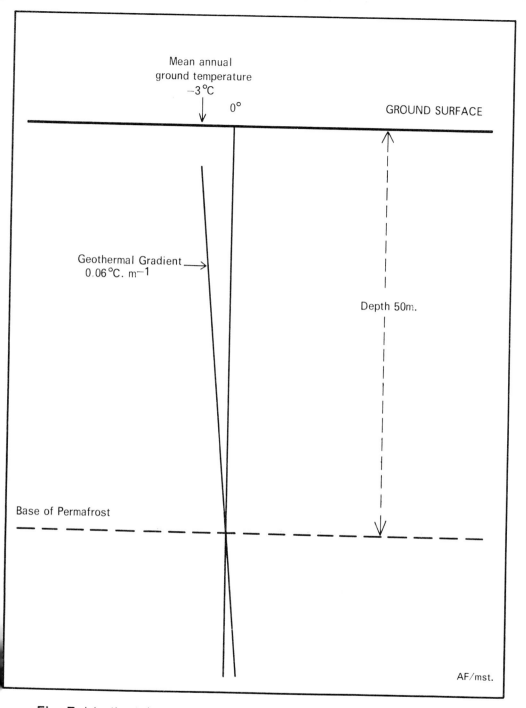

Fig. 7 Idealised (steady state) depth-temperature relationship for permafrost.

Simple analytical methods necessarily involve assumptions and approximations acceptable only for a limited number of situations. Today there is a great need for an understanding of the ground thermal regime under natural or semi-natural conditions. Such an understanding requires an intimate knowledge of field conditions, and at the same time of an analytical or theoretical framework to which these can be related.

REFERENCES

Beskow, G. 1935: Tjalbildningen och tjällyftningen med sarskild hansyn til vagar och jarnvagar. Stockholm, 242 pp. Statens vaginstitut. Stockholm, Meddelande 48. (Also published as Sveriges geologiska undersokning. Avh. och Uppsats. Ser. V. 375, and translated into English: Tech. Inst. North Western Univ., Evanston, Ill, November 1957).

Brown, R.J.E. 1966: The Influence of Vegetation on Permafrost. Proc. Int. Conf. Permafrost. U.S. Nat. Acad. Sci-NRC. Pupln. 1287. pp. 20-25.

Brown, R.J.E. 1970: Permafrost as an Ecological Factor in the Subarctic. Symp. Ecol. Subarct. Regions, Helsinki, 1966, pp. 129-140.

Brown, W.G. 1964: Difficulties associated with predicting Depth of Freeze or Thaw. Can. Geot. Jour. 1 (4), pp. 215-226.

Gold, L.W. 1963: Influence of the Snow Cover on the average Annual Ground Temperature at Ottawa, Canada. Int. Ass. Sci. Hydrol., Publn. 61, pp. 20-25.

Gold, 1967: Influence of Surface Conditions on Ground Temperatures, Can. J. Earth Sci. 4, pp. 199-208.

Grober, H., S. Erk and U. Grigull 1961: Fundamentals of Heat Transfer, McGraw Hill, 527 pp.

Janson, Lars-Erik, 1964: Frost Penetration in Sandy Soil. Trans. Roy. Inst. Technol. Stockholm, Nr. 231, Civ. Eng. 10, 167 pp.

Klove, K. 1971: Forenklete metoder for beregning av teledybder. Frost i Jord, 2, p. 15-22.

Pearce, D.C., and L.W. Gold 1959: Observations of Ground Temperature and Heat Flow at Ottawa, Canada. Journal Geophys. Res. 64, (9), p. 1293-1298.

Penner, E. 1962: Thermal Conductivity of Saturated Leda Clay. Geotechnique, XII, (2), pp. 168-175.

Penner, E. 1970: Thermal Conductivity of Frozen Soils. Can. J. Earth Sci. 7 (3), p. 982-987.

Penrod, E.B., W.W. Walton and D.V. Terrell 1958: A Method to describe Soil Temperature Variation. Jour. Soil Mechs. Found. Div., Proc. A.S.C.E. 84, SMI.

Raumbaut, A.A. 1951: Underground Temperatures at Oxford as determined by Platinum Resistance. Therm. Radcliffe Obs. Met. Obsns. Oxford, 51, 101-204

Williams, G.P. 1968: The Thermal Regime of a Sphagnum Peat Bog. Proc. 3rd Int. Peat Congr. p. 195-200.

Williams, P.J. 1961: Climatic Factors Controlling the Distribution of certain Frozen Ground Phenomena. Geogr. Ann. XLIII, (3-4) pp. 399-347.

Williams, P.J. 1964a: Experimental Determination of Apparent Specific Heats of Frozen Soils. Geotechnique XIV (2), pp. 133-142.

Williams, P.J. 1964b: Unfrozen Water Content of Frozen Soils and Soil Moisture Suction. Geotechnique, XIV (3), pp. 231-246.

THE NATURE AND USE OF TILL FABRICS

Stuart A. Harris

The fabric of a sediment is "the observation in space of the elements of which it is composed" (American Geological Institute, 1959). Since a till is composed of a wide range of grain sizes, the overall fabric of a till needs to be described in terms of the orientation of all the material from cobble size to fine clay. This would require a number of different techniques and would be very difficult to carry out on an appreciable number of samples. Fortunately Richter (1932) demonstrated that the pebbles and cobbles in a till are often found with their long axes orientated in preferred directions related to the direction of ice movement. Since it is relatively easy to measure the direction of the long axes of cobbles, it has become the practice to talk about "till fabrics" synonymously with pebble orientation in the till. Except where otherwise stated, the term "till fabrics" will be used in this sense in the following discussion.

Till fabrics are proving most valuable in studies of ice movement especially in areas of multiple till layers. In the ensuing paper, an attempt will be made to discuss the nature and use of till fabrics. Particular emphasis will be placed on some of the problems which have arisen in their determination and use.

Field Work

Till fabrics may be determined in the field, or else an orientated sample may be brought back to the laboratory for study. Field measurement is more usual and saves losing accuracy through the problem of correctly marking the field orientation on a block of till and transporting it back to the laboratory intact. On the other hand the laboratory method does save time in the field.

The field method involves carefully scraping away the finer material from a smooth surface of undisturbed till with a knife until a bladed or rod-shaped pebble is found. Removal of part of the matrix enables either the direction and dip of the long axis of the pebble or the direction of the most elongate part of the pebble in a given plane (horizontal or vertical) to be measured. Field notes should include both orientation and angle of dip of each pebble.

Alternatively orientated till cores and hand specimens may be used for rapid microscopic fabric studies (Dreimanis, 1959). The blocks are dried, and then cut with a diamond saw into slices 2-5 cms.

thick. Brushing under a water spray gradually removes the matrix exposing the larger clastic elements, whose orientation can then be determined.

If only a few pebbles greater than 2 mm. diameter can be found in even a large volume of material, then it is necessary to take an orientated block and to work on it in the laboratory. Techniques include cutting thin sections of the material for study of the fabric of the sand fraction under the microscope. Details will be found in Dapples and Rominger (1945) and Griffiths and Rosenfeld (1953).

Representation of the Results

There are two normal forms of visually presenting the data, viz., two dimensional rose diagrams and three dimensional stereographic projections. Undoubtedly the most accurate method of representing the results is the three dimensional method (Fig. 1). The poles of the long (a) axes of the pebbles are plotted on a polar stereographic projection. Since this is an "equal area" projection, i.e. each degree square is of the same size, the distributions of poles can then be contoured by one of the conventional techniques (Flinn, 1958; Kamb, 1959). However, bearing in mind that the probable accuracy of measurement of the correct angle of azimuth of a pebble may be ±5° (see Andrews and Shimizu, 1966; Hill, 1968), this method may give a false impression of accuracy. On the other hand the errors of measurement of angles will tend to cancel one another out, especially when large numbers of observations are averaged.

Three dimensional projections are poor for use on maps where a clear visual presentation of data from a number of sites is vital. For this purpose rose diagrams (usually based on 20° classes) give the best results (Fig. 2).

If all that is needed is the modal azimuth of the long axes of the pebbles, then it will be easier to arrange the data by one degree classes and then pick the mode based on, say, a twenty degree class. The centre angle of this modal class will be the modal azimuth. Krumbein (1939) has shown that the mean and mode lie within one or two degrees of one another. Another common method is to simply calculate the mean direction of the mode.

Typical Results

Most fabrics show two modes, the larger of which is usually orientated parallel to the direction of ice movement while the

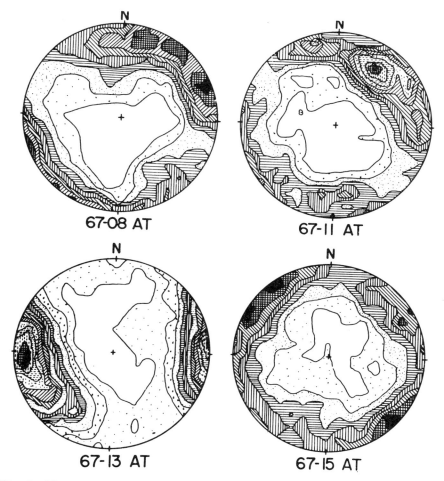

Fig. 1 Three dimensional fabrics of four samples of Adams till, Adams Inlet, Southeast Alaska (After McKenzie, 1970)

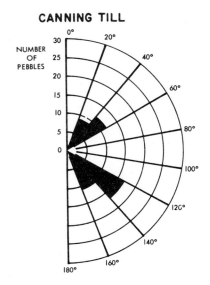

Fig. 2 Typical diagram of a till fabric (Canning till M1)

second lies at right angles to it in a horizontal plane (see Fig. 2). Cases where the larger mode lies at right angles to the direction of movement ("cross fabrics") are usually present locally in a given till, and may occasionally be the norm. There is usually a tendency for the pebbles to plunge towards the direction from which the ice was moving as in Fig. 1 (see Harrison, 1957; Harris, 1968), but this is not always the case (Glenn, Donner and West, 1957; Saunders, 1968 and Young, 1969). The modes of the fabric are not usually completely symmetrical (Ramsden, 1970), e.g. in Fig. 1.

Krumbein (1939) reported a fabric with a single mode, but the writer has only come across one of these in the course of determining over 500 fabrics. Similarly Holmes (1941) obtained a third vertical mode by averaging a number of fabrics but he could not explain its origin satisfactorily. Harris (1969) pointed out that upturning of some of the pebbles by frost action would result in a third mode of this type. Sometimes no preferred orientation is apparent.

Variability of the till both within one section and between sections is a basic feature of the deposit and is proving very important in the determination of the depositional environment (Young 1969). It is therefore most important that this aspect of the nature of tills be taken account of during the planning of a sampling programme.

Statistical Treatment of the Results

In the statistical treatment of the results of the field work, three basic questions must be answered (Harris 1969). (1) Is there a preferred orientation? (2) If so, what is the strength of the fabric in this direction? (3) What is the probability that the preferred orientation is real and not imaginary. The treatment may involve the data obtained from one site, or it may involve the study of variations of the fabric over a given area occupied by the particular till.

Although many ways have been developed for treating the data, none are entirely satisfactory. Krumbein (1939) used an arithmetic method to study the reproducibility and significance of the data from a given source. It only satisfactorily answered the question of direction of ice movement and has the disadvantage that the results are dependent on the choice of the source used as the base for the statistical work. Curray (1956) used vector analysis to measure the significance of the primary mode while Harrison (1957) used a Tukey chi-square test. Scheidegger (1965) and Fisher (1953) have tried yet other methods. For further details of the methods, see

Andrews and Smith (1970). All five methods are designed to work with unimodal data, whereas most till fabrics are bimodal.

The chi-square method has been used to overcome the problem of bimodal distributions (Kauranne, 1960; Andrews, 1963; Andrews and Smithson, (1965) but it primarily concentrates on producing parameters measuring the significance of the data. It has been criticised on the grounds that it fails to determine the directions and planes of the distribution of orientations (Ramsden, 1971). Harris (1969) suggested using a graphical technique based on the chi-square for obtaining the degree of dispersion of the primary mode at the 95% level of probability (the minimum significant orientation count or MSOC). This value can then be plotted on maps or compared to variations in possible causative factors, e.g., roughness of the land, speed of ice movement, clast size, amount of clay present, etc.. A two-dimensional method by Jones (1968) looks promising, but unfortunately it has yet to be applied to tills.

A further development was the concept of rotation of the data so as to enhance the size of the primary mode (Andrews & Shimizu, 1966). This originated with work on paleomagnetic vectors (Watson and Irving, 1957; Steinmetz, 1962) and computer programmes have now been developed to select the optimum rotation (Mark, 1971). The question which arises here is just what does this produce? Can the product be related to depositional conditions or does the statistical manipulation render environmental correlations either suspect or false? These questions remain unanswered, though it is assumed that the improved clustering more nearly corresponds to the conditions of deposition.

A common use of till fabrics is to plot them on maps in order to indicate the probable flow direction of the parent ice sheet (see West and Donner, 1956). To these plots may be added probable flow lines (e.g. Harris, 1968; 1969; 1970), and Roberts and Mark (1970) have now described a means of obtaining these probable flow lines by computer.

Interpretation and Use of the Results

The presence or absence of well developed modes, the azimuth of the primary mode, and the variability of the fabric laterally and vertically within a given till are all properties which can be used in identifying the nature of the depositional environment. Thus active ice moving at normal rates produces consistent primary modes orientated parallel to the direction of the ice movement, while surging of a glacier appears to produce a till characterised by transverse fabrics (Rutter, 1969). Ice stagnation results in a considerable amount of modification of the fabrics during the melting process with the result that the fabrics either show no significant modes, or else the primary modes lie orientated randomly. Naturally, if the direction of ice movement changes during the deposition of a given till, the upper layers will have a different fabric direction to the lower layers.

Degree of dispersion about the primary mode may also be significant, although confusion currently exists regarding its implications. Both Holmes (1941) and Andrews and King (1968) have found evidence that variations in pebble shape produce marked differences in the till fabric. Lindsay (1970) encountered contrary evidence from a Permian tillite in Antarctica. According to Lindsay, increasing pebble size was strongly correlated with greater dispersion while thicker deposits are the result of more rapid deposition and retain better fabric orientation. Clay content also correlated with better orientation, while pebble lithology was also important. In the same year Ramsden (1970) concluded that coarser clasts were strongly correlated with stronger fabrics based on a study of tills near Edmonton!

These discrepancies between evidence from different places indicate the need for caution in obtaining evidence from one area and applying it elsewhere. The experience of the writer in Ontario strongly suggested a lack of correspondence between clast size or clay content and degree of dispersion of the fabric (see Harris, 1969). The only correlation that was obvious was with roughness of the terrain. A test of fabric dispersion versus rate of movement in ice in Colorado suggested that there might be a correlation with speed of movement of ice (Harris, 1968) but it remains to be seen whether this proves to be correct. There would appear to be a number of possible factors involved in controlling the degree of dispersion, but much more work is needed before we know what the causes of the observed variations may be.

There is also considerable doubt as to the cause of the few scattered transverse fabrics resulting from active ice. The work of Rutter (1970) would suggest that they are the result of deposition during local surging of the ice sheet. Andrews and King (1968) thought that when passing over a rough surface, localized compression would cause transverse fabrics to be produced whereas localized tension would produce fabrics trending parallel to the direction of movement. Glen, Donner and West (1957) suggested that transverse fabrics were the products of long distance transport whereas the other fabrics were produced by movement over short distances. Since the latter commonly occur in Laurentide tills in Alberta, as well as at the margins of cirque glaciers, this seems rather an unlikely explanation. Holmes (1941) explained transverse fabrics as being produced by a predominance of pebbles rolling beneath the ice. Pebbles lying with their long axes parallel to the direction of movement were regarded as being the product of sliding beneath the ice. Introduction of material with a different fabric into the till by erosion is another possibility. Unless there was rapid re-orientation, the resultant till fabric would differ while thicker deposits are the result of more rapid deposition and retain better fabric.

Deformation Fabrics

Between the time a given till is deposited and the time when it is sampled, it may be modified by various agents. A clear understanding of the various types of deformation which may take place is vital to a reliable study of till fabrics. The agencies which can cause deformation are readvance of an ice sheet, frost action, soil creep and ice pressing. Soil formation without appreciable disturbance of the layers by animals produces little modification other than by weathering of the less resistant pebbles (see Table I, after Harris, 1969).

When an ice sheet advances over a soft unconsolidated and unfrozen till, tight folds may be produced in the underlying material (Banham, 1968; Ramsden & Westgate, 1971). This produces a rotation of the pebbles in the till in the limbs of the fold so that they come to lie with their long axes at right angles to the fold axis. In the axial region of the fold, the pebbles become aligned parallel to the fold axis. This phenomenon has rarely been observed and described although it would cause a serious modification of the original fabric and is yet another potential cause of cross fabrics.

At sites where the till is subjected to marked freezing and thawing, the pebbles in the upper metre of a deposit will first become rotated until the long axis is vertical, and then they will move upwards through the soil (Harris, 1969). Where another deposit such as aeolian sand or lake silts overlie the till, the pebbles will move upwards into the surface deposit. In the four cases described from Ontario, the fabrics always retained their general direction of orientation even when the pebbles had migrated into the new deposit (Fig. 3). This is therefore another possible way in which a

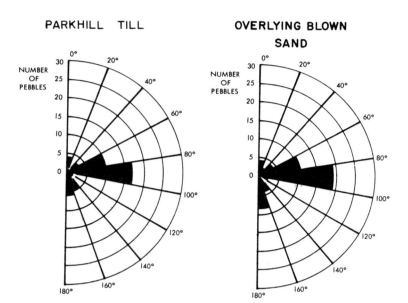

Fig. 3 Deformation fabric in blown sand (E6) and the apposition fabric in the underlying Parkhill till (E3) at G.R.352, 127, west of Waterloo, Ontario (after Harris, 1969)

TABLE I

REPLICATE DETERMINATIONS OF PEBBLE ORIENTATION IN THE KANSAS TILL
(after Harris, 1969)

Sample Number	Soil horizon	Based on 50 pebbles		
		Pebbles in primary mode	MSOC	Direction of primary mode
LR22	C_g	17	60	N153°
LR24	B_2	19	49	N152°
LR34	CG	19	49	N153°
LR41	C_g	22	38	N155°
LR42	A_1/A_2	20	44	N155°
LR43	B_2	18	53	N156°
Mean	–	19	49	N154°
Standard Deviation	–	1.34	–	1.56

53

transverse fabric can be produced in a thin surface till. It also means that the presence of a till-like fabric in a thin sediment is a poor indicator of the origin of the sediment. Identification of depositional environment must involve other methods such as grain size analysis, presence or absence of bedding, etc..

Soil creep is ubiquitous in almost all temperate areas except perhaps where water or wind erosion is extreme. The upper layer of the soil tends to move slowly downhill. In Southern Ontario, the rate is probably fairly slow (see Everett, 1963, who found a mean rate of 0.5mm/year in Ohio). A study of the effect of this movement in the two cases reported so far (Harris, 1969) shows that in this environment, the fabric of the parent deposit is destroyed, and is replaced by a random orientation (Fig. 4). Thus layers which have been affected by soil creep must be omitted from a study of till fabrics.

Recent work had indicated that the effect of more rapid soil creep is different from that in Ontario. Research being carried out at the Kananaskis Environmental Science Centre in Alberta indicates that slope movement may be as high as 150 mm in a month at some sites in the outer ranges of the Rocky Mountains (Harris, 1971). Fabrics in the colluvial material that is derived from till show a marked down-slope primary mode (Fig. 5) which is at variance with the primary mode in the underlying till (Fig. 6). Since the layer involved in this surface movement is thick and occurs over most of the lower mountain slopes, great care is needed in tracing ice movements in the area. Only fabrics not trending straight down-slope can be used. Since no sorting takes place during soil creep, the layer which is moving cannot be readily distinguished from the underlying till except in this way.

It is also probable that ice pressing of soft till beneath a melting ice sheet produces changes in the fabric, but so far no work appears to have been published on this problem. Some field work has just been commenced near Calgary to try to remedy this deficiency.

Distinguishing Till from other Diamictons

So far, we have discussed the separation of till from colluvium. There are a number of other diamictons which may be present in areas where till may occur and which must be distinguished from till if a correct paleogeographical interpretation is to be made.

In glaciated lowlands, the main diamictons which may cause confusion other than colluvium are slides, mudflows and ice-rafted

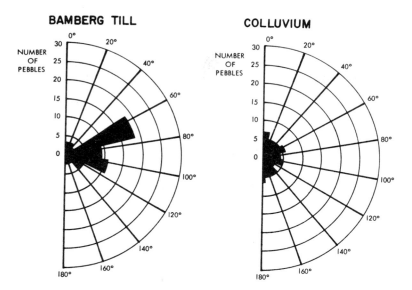

Fig. 4 Fabric for the Bamberg till and overlying colluvium at G.R. 323, 123, near Erbsville, Ontario.

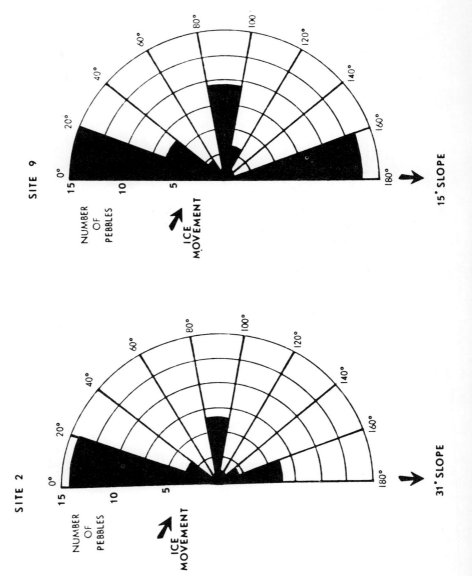

Fig. 5 Preferred orientation of pebbles in a downslope direction in colluvium overlying till on the mountainside north of Barrier Lake, Kananaskis, Alberta. The direction of ice movement is based on fabrics from the underlying till (see Fig. 6).

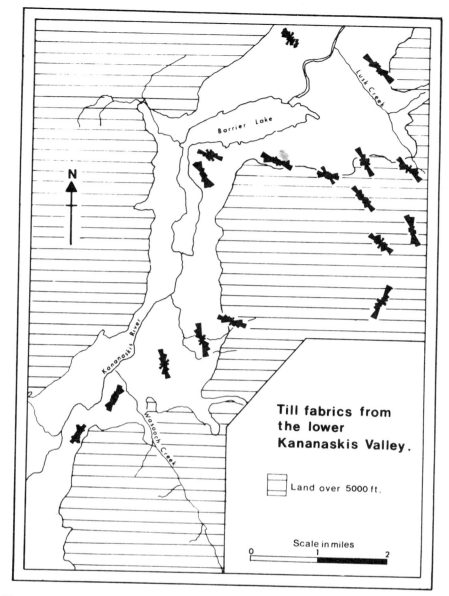

Till fabrics from
the lower
Kananaskis Valley.

▭ Land over 5000 ft.

Scale in miles
0   1   2

Fig. 6   Till fabrics for the last ice advance into the Lower Kananaskis
Valley, after Walker, 1971.

deposits. The products of slides normally have a random orientation, as do ice-rafted deposits. The fabrics, taken together with other properties such as lithology, presence of ostracodes, etc., make these deposits readily recognisable. Mudflows are more difficult in that they produce a similar fabric to a till deposited by an active ice sheet (Lindsay, 1968). While there may be small differences between the fabrics, we obviously have too little data at the moment to safely distinguish the deposits on the basis of fabrics alone. In lowland areas, this is no great problem due to the normal disparity in size and extent of the two phenomena.

In mountain regions, this can be another matter. Both glaciers and mudflows tend to follow drainage ways. Glaciers leave lateral and ground moraines while mudflows leave mudflow levees and extensive deposits. Fortunately mudflows were less extensive and frequent than glaciers were during the Pleistocene, although the risk of confusion is obvious. This problem reaches its peak when diamictons are studied high on a volcanic mountain in the humid tropics. Hot lava and ash flows can produce a deposit with boulders in it. These flows follow the valleys, just as mudflows and glaciers do. All three processes can produce slickensides on the underlying deposits. Weathering in the warm humid summers rapidly destroys the fine grained material producing a matrix of clays for the boulders. Thus all three deposits tend to look alike (A.M. Neumann, personal communication). Fortunately the boulders from the hot ash flows tend to have a random orientation, while the mudflows (lahars) tend to exhibit some sorting along their length. More work is needed on the differentiation of these deposits because it makes a big difference to palaeographic interpretations as to which origin a given deposit is assigned!

Other diamictons in mountain regions such as screes and rock glaciers can be readily distinguished by other properties such as the slope, texture and form. Thus mudflows and colluvium are the main problems.

Use of Till Fabrics

Apart from the recognition of past deposits, till fabrics are proving a tremendous boon in determining the direction of ice movement in the past. Ice sheets and glaciers do not often leave obvious end-moraines. Another method of determining direction of ice movement is by using the direction of elongation of drumlins. The main problem here is that drumlins may be formed by erosion or deposition, and in cases of more than one ice advance into an area, it becomes difficult to be certain as to which of the ice advances

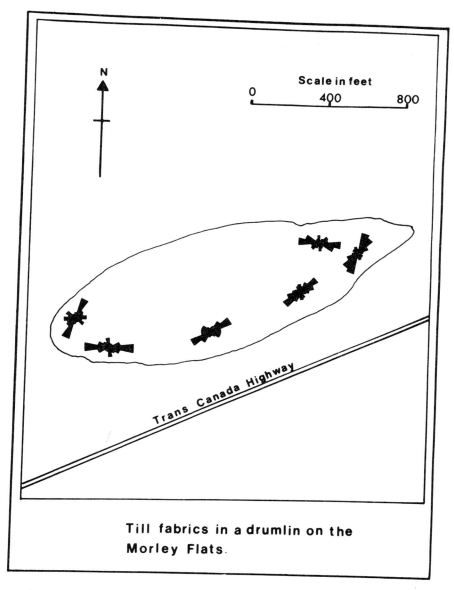

Till fabrics in a drumlin on the Morley Flats.

Fig. 7   Fabrics on the margin of a drumlin formed by deposition from the Bow Valley advance, Morley Flats, Alberta (after Walker, 1971)

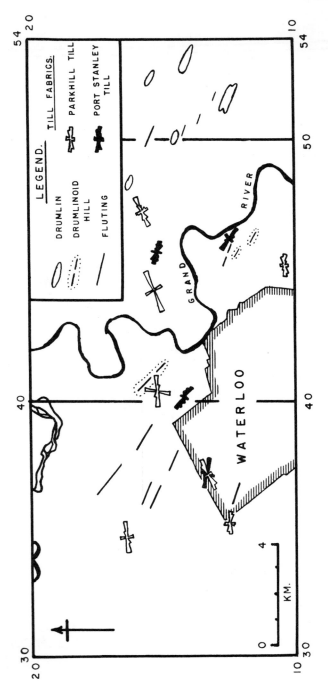

Fig. 8 Direction of elongation of drumlins, drumlinoid features, and glacial flutings compared with the Parkhill and Port Stanley till fabrics near Waterloo, Ontario. Note that the earlier till fabric (Port Stanley) is the one which appears to match the drumlin direction best. The material of which the drumlins are composed includes older tills and outwash deposits, i.e., the drumlins are erosional in origin.

the drumlins are related. Thus the drumlins themselves increase the need for an independent method of determining the direction of ice movement.

By using till fabrics, the puzzle can often be solved. Walker (1971) has illustrated what happens when a drumlin is formed by deposition from an ice sheet (Fig. 7). The ice bends slightly as it moves past the hill but the general fabric direction is parallel to the direction of elongation of the drumlin. In the case of drumlins of erosional origin, it is necessary to contrast the direction of their long axes with the fabric directions of the tills nearby (Fig. 8). Thus in the case of the drumlins east of Waterloo, Ontario, the erosion of the drumlins appears to have been completed during the Port Stanley ice advance.

By putting together the information obtained by determining till fabrics at sites within two miles apart, the patterns of movement of an ice sheet can be determined. By doing this it has been discovered that the North American ice sheet advanced as numerous lobes (e.g. Wright, 1957; 1962, in Minnesota; Alley, 1971, in Alberta; and Harris, 1970 in Ontario). This has led Wright (1970) to question whether the margin of the Laurentide ice sheet did, in fact, move back and forth everywhere around its margin at the same time. The importance of this in correlations of Pleistocene stratigraphy is obvious.

CONCLUSIONS

It is obvious from the foregoing account that a great amount of work needs to be carried out before we can claim to have answered all the questions associated with the nature of till fabrics. However it is also clear that the development of this technique represents one of the major break-throughs of the last half century in the study of glacial deposits, and it should well repay continued work.

ACKNOWLEDGEMENTS

The writer is indebted to P. Ojamaa who kindly criticised this manuscript, and to M.J. Walker for permission to use Fig. 6 and 7.

REFERENCES

Alley, N.F., 1971, *Pleistocene Ice Movements and Glacial Limits West of the Porcupine Hills, Southwestern Alberta.* Abstracts, Geol. Soc. Amer., Rocky Mtn. Section, Calgary, pp. 365-366.

American Geological Institute, 1957, *Glossary of Geology and Related Sciences.* American Geological Institute, Washington, 325 pp.

Andrews, J.T., 1963 *The cross-valley moraines of North central Baffin Island: A quantitative analysis.* Geog. Bull., No. 20, pp. 82-129.

Andrews, J.T., and King, C.A.M., 1968, *Comparative till fabrics and till fabric variability in a till sheet and a drumlin: a small-scale study.* Proc. Yorks. Geol. Soc., v. 36, pp. 435-461.

Andrews, J.T., and Shimizu, K., 1966, *Three dimensional vector technique for analyzing till fabrics: Discussion and Fortran Program.* Geographical Bulletin, v. 8, pp. 151-165.

Andrews, J.T., and Smith D.I., 1970. *Statistical Analysis of Till Fabric: Methodology, Local and Regional Variability.* Q.J. Geol. Soc., London, v. 125, pp. 503-542.

Andrews, J.T. and Smithson, B.B., 1965. *Till Fabrics of the Cross-Valley Moraines of North-central Baffin Island, Northwest Territories, Canada.* Geol. Soc. Amer. Bull., v. 77, pp. 271-290.

Banham, P.H., 1966. *The significance of Till Pebble Lineations and Their Relation to Folds in Two Pleistocene Tills at Mundesley, Norfolk.* Proc. Geol. Assoc., London, v. 77, pp. 469-474.

Curray, J.R., 1956. *The Analysis of Two-dimensional Orientation Data.* J. Geol., v. 64. pp. 117-131.

Dapples, E.C. and Rominger, J.F., 1945. *Orientation Analysis of Fine-Grained Clastic Sediments: A Report of Progress.* Jour. Geol., Vol. 43, pp. 246-61.

Dreimanis, A., 1959. *Rapid Macroscopic Fabric studies in drill-cores and Hand specimens of Till and Tillite.* J. Sed. Pet., v. 29, pp. 459-463.

Everett, K.R., 1963. *Slope Movement, Neotomia Valley, Southern Ohio.* Inst. Polar Studies, Report No. 6, Columbus, Ohio, 62 pp.

Fisher, R.A., 1953. *Dispersion on a sphere.* Proc. Roy. Soc., London, Ser. A., v. 217, pp. 295-306.

Flinn, D., 1958. *On Tests of Significance of Preferred Orientation in Three-dimensional Fabric Diagrams.* J. Geol., v. 66, pp. 526-539.

Glen, J.W., Donner, J.J., and West, R.G., 1957. *On the Mechansim by which Stones in Till become Orientated.* Am. J. Sci: v. 255, pp. 194-205.

Griffiths, J.C., and Rosenfeld, M.S., 1953. *A further test of dimensional orientation of quartz grains in Bradford Sand.* Amer. J. Sci., v. 251, pp. 192-214.

Harris, S.A., 1968. *Till fabrics and speed of movement of the Arapahoe Glacier, Colorado.* The Prof. Geog., v. 20, pp. 195-198.

Harris, S.A., 1969. *The Meaning of Till Fabrics.* Canadian Geographer, v. 13, pp. 317-337.

Harris, S.A., 1970. *The Waterloo Kame moraine, Ontario, and its Relationship to the Wisconsin Advances of the Erie and Simcoe ice lakes.* Zeitschrift fur Geomorphologie, v. 14, pp. 487-509.

Harris, S.A., 1971. *Preliminary Observations on Downslope Movement of Soil during the Fall at Kananaskis, Alberta.* Abstracts, Geol., Soc. Amer., Rocky Mtn. Section, Calgary, pp. 385-386. also in Proc. Can. Assoc. Geogr., Waterloo, Ontario.

Harrison, P.W., 1957. *A Clay Till Fabric: Its Character and Origin.* J. Geol., v. 65, pp. 275-308.

Hill, A.R., 1968. *An Experimental Test of the Field Techniques of Till Macrofabric Analysis.* Trans. Inst. Brit. Geog., v. 45, pp. 93-105.

Holmes, C.D., 1952. *Drift Dispersion in West-Central New York.* Geol., Soc. Amer. Bull., v. 63, pp. 993-1010.

Kamb, W.B., 1959. *Ice Petrofabric Observations from Blue Glacier, Wash., in Relation to Theory and Experiment.* J. Geophys. Res., v. 64, pp. 1891-1909.

Kauranne, L.K., 1960. *A Statistical Study of Stone Orientation in Glacial Till.* Finland, Comm. Geol. Bull., No. 188, pp. 87-97.

Krumbein, W.C., 1939. *Preferred Orientation of Pebbles in Sedimentary Deposits.* J. Geol., v. 47, pp. 673-706.

Lindsay, J.F., 1968. *The Development of Clastic Fabric in Mudflows.* J. Sed. Pet., v. 38, pp. 1243-1253.

Lindsay, J.F., 1970. *Clastic Fabric Strength of Tillite.* Jour. Geol., v. 78, pp. 597-603.

Mark. D.M., 1971. *A Rotational Vector Procedure for the Analysis of Till Fabrics.* Geol. Soc. Amer. Bull., In the Press.

McKenzie, G.D., 1970. *Glacial Geology of Adams Inlet, Southeast Alaska.* Institute of Polar Studies. Report No. 25, Columbus, Ohio, 121 pp.

Ramsden, J., 1970. *Till Fabric Studies of the Edmonton Area.* Unpublished M.Sc. Thesis, Dept. of Geology, University of Alberta.

Ramsden, John, 1971. *Three Dimensional Till Fabric Analysis-Alternative Approaches.* Abstracts, G.S.A.; Rocky Mtn. Section, Calgary, Alta., p. 406.

Ramsden, J., and Westgate, J.A., 1971. *Evidence for Reorientation of a Till Fabric in the Edmonton area, Alberta.* In R.P. Goldthwait (ed.) Symposium on Till, Ohio State University Press. In the press.

Richter, K. 1932. *Die Bewegungsrichtung des Inlandeises rekonstruiart aus don Kritzen und Lungsachsen der Geschiebe.* Zeits f. Geschiebeforschung, v. 8, pp. 62-66.

Roberts, M.C., and Mark, D.M., 1970. *The Use of Trend Surfaces in Till Fabric analyses.* Can. J. Earth Sci., v. 7, pp. 1179-1184.

Rutter, N.W., 1969. *Comparison of moraines formed by surging and normal glaciers.* Can. J. Earth Sci., v. 6, pp. 991-999.

Saunders, G.E., 1968. *A fabric analysis of the ground moraine deposits of the Lleyn Peninsula of south-west Caernarvonshire.* Geol. J., v. 6, pp. 105-108.

Scheidegger, A.E., 1965, *On the Statistics of the Orientation of Bedding Planes, Grain axes, and Similar sedimentological data.* U.S.G.S. Prof. Paper 525-c, pp. 164-167.

Steinmetz, R., 1962. *Analysis of Vectoral Data.* J. Sed. Pet., v. 32, p. 801-812.

Walker, M.J., 1971. *Late-Wisconsin Ice in the Morley Flats Area of the Bow Valley and adjacent areas of the Kananaskis Valley, Alberta.* Unpublished

M.Sc. Thesis, Dept. of Geography, University of Calgary. Also in abstracts, Geol. Soc. Amer., Rocky Mtn. Section, Calgary, p. 418.

Watson, G.S., and Irving, E., 1957. *Statistical Methods in Rock Magnetism.* Mon. Nat. Roy. Astron., Soc., Geophysical Supplement, v. 7, pp. 289-300.

West, R.G., and Donner, J.J., 1956. *The Glaciations of East Anglia and the East Midlands: a differentiation based on stone orientation Measurements of the Tills.* Q.J. Geol. Soc. Lond., v. 112, pp. 69-91.

Wright, H.E., 1957., *Stone Orientation in Wadena Drumlin Field, Minnesota.* Geografiska Annaler, v. 39, pp. 19-31.

Wright, H.E., 1962. *Role of the Wadena Lake in the Wisconsin Glaciation of Minnesota.* Bull. Geol. Soc. Amer., v. 73, pp. 73-100.

Wright, H.E., 1970. *Retreat of the Laurentide Ice Sheet from 14,000 to 9,000 year ago.* Abstracts, AMQUA meeting, Bozeman, Montana, pp. 157-159.

Young, J.A.T., 1969. *Variations in Till Fabric over very short distances.* Bull. Geol. Soc. Amer., v. 80, pp. 2343-2352.

THE EFFECT OF LITHOLOGY UPON TEXTURE OF TILL

A. Dreimanis and U.J. Vagners

INTRODUCTION

Texture of till depends upon several factors, most important being the following four:

(1) lithologic composition of both matrix and clasts;
(2) comminution and possible sorting of rock and mineral fragments during the transport of drift by glacial ice and its meltwaters prior to desposition of till; the comminution depends greatly upon the mode of glacial transport - superglacial, englacial, or basal;
(3) process of deposition, for instance by lodgment, basal melting, ablation, etc. - see Dreimanis (1969, Tab. 1) or Dreimanis and Vagners (1971 Fig. 2) for classification of tills in relation to transport and deposition of drift material;
(4) postdepositional changes.

This discussion will consider two factors only: lithology and comminution during glacial transport by an ice sheet - where the transport is mainly basal or englacial. The till samples investigated were taken from southern and central Ontario (Fig. 1) thus making it possible to consider a variety of lithologies, as bedrock of this area consists of Precambrian igneous and metamorphic rocks in the north and Paleozoic nonmetamorphic sedimentary rocks in the south. Regional glacial movement has been mainly from the north, or north-east, with additional lobal movements radiating out of the Great Lakes' depressions in southern Ontario.

Field examination suggests that most samples were from basal till. In order to avoid excessive amounts of local material incorporated in the tills studied, the sampling was done at least 0.6 m above the base of till. (Still two of the samples turned out to be local tills: No. 16 and one in the traverse III.) Nonoxidized material was preferred, and in no case leached or even partly leached till, or till containing visible secondary minerals was taken. The granulometric analyses were done by sieving and hydrometer method, and the Wentworth scale was applied for establishing boundaries between various particle size grades (Wentworth, 1922). In the first 25 large samples (1 cb.m average) the particle sizes ranged from 25.6 cm to less than 2 μ in others (Vagners, 1969) from 32 mm to less than 1 μ. The lithologic composition of each Wentworth grade was determined

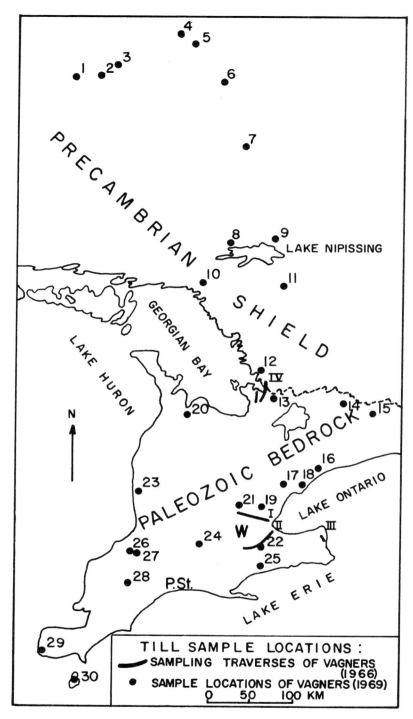

Fig. 1 Locations of till samples discussed in this paper: dots-individual samples, heavy lines-samples along traverses, P.St.-area of Port Stanley Till, W-area of Wentworth till.

separately. Detailed descriptions of sampling sites, laboratory procedures, and the results of analyses of individual samples are in Vagners (1966 and 1969). Typical results of analyses, selected from the above 55 sites of investigation will be used for illustrating this discussion, by using mainly frequency polygons, histograms and cumulative curves; the Wentworth grade scale is shown in milimeters and in its logarithmic transformation, the phi units (Krumbein, 1934).

Most published analyses of tills deal with the texture of till matrix only, with 2 mm or 4 mm as its upper boundary. The clasts of boulder, cobble and pebble sizes are usually excluded, mainly for practical reasons: determination of all the particle sizes present in till would require extremely large samples. Because of the exclusion of clasts (pebbles, cobbles, boulders), the usual terms "texture of till" or "granulometric composition of till" should be supplemented by the word "matrix" added after "till". It should be realized also that cummulative granulometric curves of the same till will look quite different, depending upon the coarsest particle size chosen (Fig. 2).

Bimodal distribution of lithic components

Texture of the entire till, including not only matrix, but also clasts, is governed mainly by the rules of bimodal distribution of its lithic components, minerals and rocks (Dreimanis and Vagners, 1969 and 1971). In the simplest case, if a till consists of fragments of a monomineralic rock only, for instance dolostone, and its mineral dolomite, the rock fragments are represented by at least one mode in the particle size grades, coarser than 0.1 mm, while its mineral grains cluster in another mode in the coarse silt size (Fig. 3). The more the rock (dolostone) is comminuted, the relatively larger becomes the mineral mode: in Fig. 3 it is obvious that the mineral mode increases on the account of the rock mode(s) with increasing distance of glacial transport.

As till usually consists of several lithic components, its granulometric composition is very often multimodal (Figures 4 and 5). Even though these illustrations represent only a part of tills with 25.6 cm and 3.2 cm as the upper boundaries of the particle sizes investigated, the multimodal nature of till and its dependence upon the modes of different lithic components is obvious.

If averages of several cumulative curves of lithologically different tills are plotted either on the Rosin and Rammler's "law of crushing" paper (for instance the composite of 38 samples of

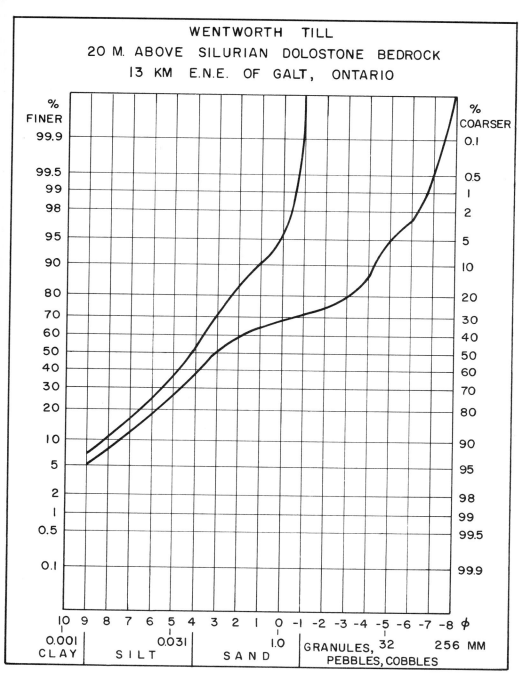

Fig. 2 Cumulative granulometric curves of a Wentworth till sample depending upon the upper boundary chosen.

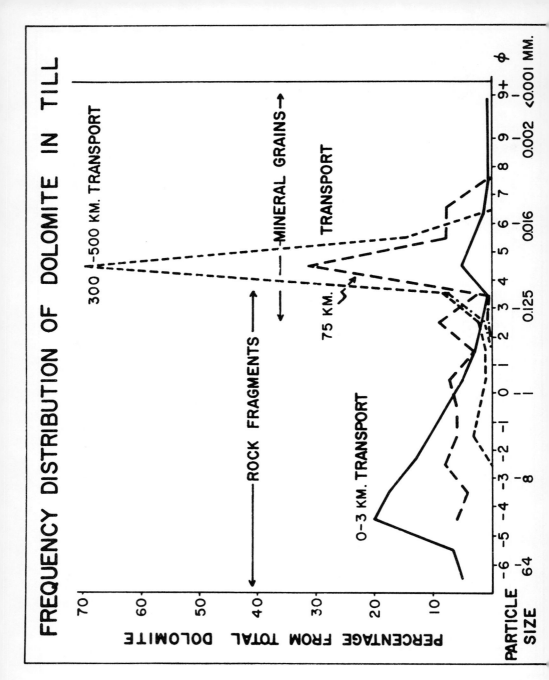

Fig. 3 Frequency distribution of dolostone-dolomite in three selected till samples from Hamilton-Niagara area (Traverses II and III, see Fig. 1). After Dreimanis and Vagners (1971 Fig. 4).

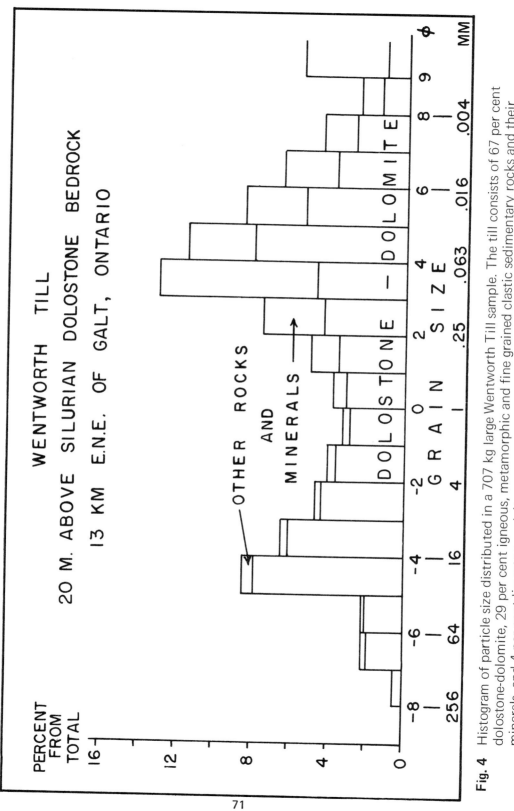

Fig. 4 Histogram of particle size distributed in a 707 kg large Wentworth Till sample. The till consists of 67 per cent dolostone-dolomite, 29 per cent igneous, metamorphic and fine grained clastic sedimentary rocks and their minerals, and 4 per cent limestone-calcite.

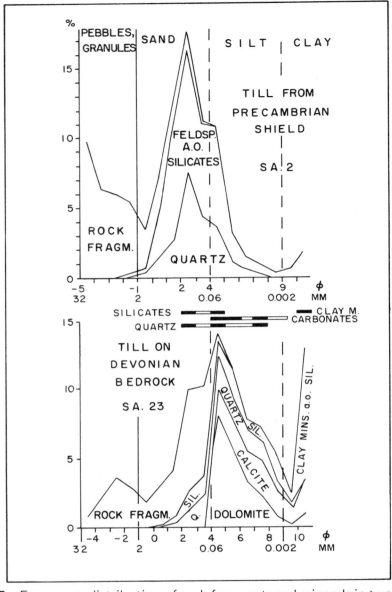

Fig. 5 Frequency distribution of rock fragments and minerals in two till samples: upper boundary — 32 mm. Sample no. 2 was collected about 50 km W. S. W. of Timmins over volcanic bedrock downglacier from an extensive area of coarse grained igneous and metamorphic bedrock; the 2.5 kg large sample consists of 42% rock fragments, 30% feldspars, 21% quartz, 3% heavy minerals, 4% fine grained silicates. Sample No. 23 was collected 5 km S.E. of Goderich in Devonian limestone bedrock area; the 1.6 kg large sample consists of 25% rock fragments 22% of each of dolomite and clay minerals, 14% calcite, 13% quartz, 4% feldspars and heavy minerals. The horizontal bars between both diagrams indicate terminal grades of most common minerals (compared with Fig. 6).

Goldthwait's lower till in Elson, 1961) or on the similar log probability paper (Adams' Cornwall till averages and grading limits in Lo and Roy, 1969 and Dreimanis' ranges of two major groups of tills in Ontario in Flint, 1971), nearly straight-line cumulative curves may be produced. They may also result, if a till consists of a variety of rocks and minerals mixed together in such proportions that their modes overlap over a wide range. Such well proportioned natural mixtures are rare, and therefore the suggestion of Elson (1961) that most tills may be recognized by straight-line curves on Rosin and Rammler's "law of crushing" paper, applies seldom to individual till analyses.

Texture of till matrix

Terminal grades of minerals. Texture of till matrix only - finer than 2 mm - depends mainly upon the terminal grades (Fig. 6) of its constituent minerals (Dreimanis and Vagners, 1969, and 1971). A terminal grade is the final product of glacial comminution, and its particle size range for each mineral depends upon the original sizes of the mineral grains while still in rocks, and upon the resistance of each mineral against comminution during the glacial transport. Comminution is accomplished by crushing, abrasion, and various other mechanical means. As the effect of abrasion decreases with decreasing particle sizes, crushing or splitting by other causes along cleavage planes and other zones of weakness inside the mineral grains and between them influences greatly the ease of comminution towards the terminal grade. Fig. 3 shows that the mineral dolomite begins to concentrate in its terminal grade already from the very beginning of its incorporation in glacial drift, 0-3 km from its source, and the particle size range of the terminal grade of dolomite remains the same over a distance of transport of several hundred kilometers. According to Vagners (1969), even such hard, but relatively brittle minerals as garnets and feldspars are comminuted to their terminal grades after glacial transport of 80-180 km, but they do not become comminuted further beyond the terminal size range even after a transport of several hundreds of kilometers.

The coarsest-grained terminal grades - in the sand and coarse silt sizes - are formed by most common igneous and metamorphic minerals: feldspars, quartz and heavy minerals (amphilbole, pyroxenes, garnets, etc.), as seen in Fig. 6. Main minerals of sedimentary rocks, except for quartz from sandstone, a rock not abundant in the area of investigation, have formed their terminal modes in finer particle sizes (Fig. 6): calcite, dolomite, and quartz from silt-stones and shales in the silt size; clay minerals were found

73

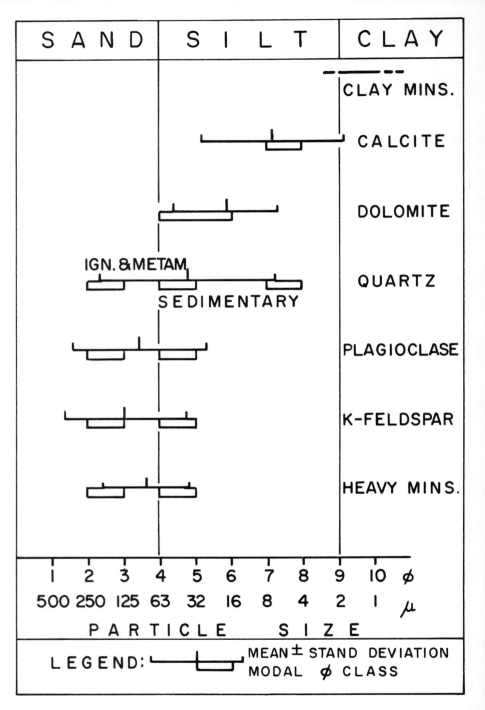

Fig. 6 Terminal grades — means, standard deviations and modal phi classes — of selected minerals in basal till. After Vagners (1969).

74

mainly in the clay size, but the range of their terminal size was not investigated. Thomas, 1969, found, that the sediments of Lake Erie which have derived largely from the tills of the adjoining land, contain their detrital clay minerals predominantly in the less than $2\,\mu$ size. Several minerals are present in two or three modal classes, for instance quartz - in three. Judging from the results of Vagners' (1969) till analyses, the coarsest and probably also the intermediate modal classes (2-3 ϕ and 4-5 ϕ) of quartz in Ontario have derived mainly from igneous and metamorphic rocks, while the finer grained quartz grains (modal classes of 7-8 ϕ and 4-5 ϕ) come from the fine-grained clastic sedimentary rocks, particularly silt-stones (see typical examples in Fig. 5). It appears, that the terminal grades of those grains of quartz which were already relatively fine, e.g. of silt size, in the original rock, are similar to their original sizes. In other words, fine quartz grains must be resistant to comminution. However, quartz grains coarser than 0.25 mm are so common in the plutonic igneous and some metamorphic rocks over the Canadian Shield, that they should be well represented in tills collected in that area, if the original particle size rather than comminution would have governed all particle sizes of quartz in tills. In the 12 till samples collected over the Precambrian terrain (Vagners, 1969), only an average of 16% of quartz (by weight) is present in all the coarse particle sizes (over 0.25 mm), while 24% is already in the 0.25-0.125 mm grade which is the coarsest of the three modal size classes of quartz.

Till matrix. The upper boundary of frequency polygons of quartz, feldspars, and heavy minerals (included in "other silicates" in Fig. 5) is close to the usual upper boundary used for granulometric till analyses on this continent: 2 mm. Therefore Dreimanis (1969) suggested that 2 mm may be considered as a natural boundary between till matrix and clasts, because most tills contain a fair to a large amount of above minerals. However, if minerals of limestone, dolostone, shale and siltstone which are the most common sedimentary rocks are considered, then the sand-silt boundary (0.06 mm) appears to be a more natural dividing line between the matrix and clasts, as it separates the terminal grades of minerals from rock fragments. For tills rich in sandstones, the clast-matrix boundary may vary, depending upon the particle sizes of the quartz derived from sandstone.

Rock fragments. Most discussions of lithologic composition of till matrix, including the recent ones by both authors (Dreimanis and Vagners, 1969 and 1971, Dreimanis, 1969), deal with minerals only, as they are main constituents of the matrix. However, if the upper boundary of till matrix is placed at 2 mm, which is commonly done when analysing tills, then also rock fragments should be considered, particularly in the coarse particle size grades. The coarsest Wentworth grade of till matrix, 2-1 mm, always contains more rocks (67-98%) than minerals. As seen in Table 1 which summarizes Vagners (1969) data, minerals already become dominant in the grades finer than 1 mm in tills consisting of igneous and metamorphic materials: these particle size fractions consist mainly of the terminal grades of various silicate minerals (compare with Fig. 6). However, in till matrix collected over the sedimentary Paleozoic bedrock, rock fragments are more abundant, and minerals become dominant only beginning with 0.25 mm, in most samples. Again, this is a function of the upper boundary of the terminal grades - in this case the terminal grades of the sedimentary minerals, even though some of the tills Nos. 13-30 contain also fair amount of igneous and metamorphic minerals.

Highest admixture of rock fragments is found in matrix of those tills which contain large amounts of medium-hard fine-textured rocks, such as shale, siltstone, limestone, slate, for instance in local tills consisting mainly of these rocks. A typical example is sample No. 16 which was taken from the area of Collingwood Shale bedrock and consists nearly entirely of fragments of this shale (Vagners, 1969). Till matrix of No. 16 contains 58% sand, 28% silt and merely 14% clay.

Predicting the texture of till matrix. When considering all the various components of till matrix in Ontario including rock fragments finer than 2 mm, a variety of cumulative granulometric curves may develop (Fig. 7). If the rock fragments are not very abundant, and the predominant minerals of till matrix are known, its texture may be predicted. For instance, most igneous and metamorphic minerals will produce a silty sand-till matrix, that is with sand as the dominant and silt as the second most common constituent (Fig. 8: I+M).

Till derived mainly from limestone and dolostone will have a predominantly silty matrix, though it will contain also an appreciable amount of sand size particles of limestone and dolostone, and some clay size carbonates. In Figure 8, the average cumulative curve (L+D) for tills Nos. 21-25 from limestone and dolostone bedrock area, is coarser than expected for a calcite and dolomite rich till, because in addition to the 37% calcite and dolomite it contains also about 30% minerals from igneous and metamorphic rocks, brought southward from the Canadian Shield by englacial transport.

TABLE 1

PERCENTAGE OF TILL SAMPLES CONTAINING MORE ROCKS THAN MINERALS IN INDIVIDUAL PARTICLE SIZE GRADES OF TILL MATRIX

(Data from Vagners, 1969)

Particle size grade	Tills collected on Precambrian Shield (samples No. 1-12)	Tills collected over Paleozoic sedimentary bedrock (samples No. 13-30)
2-1 mm	100%	100%
1-0.5 mm	30%	94%
0.5-0.25 mm	10%	55%
0.125-0.25 mm	0	38%
less than 0.125 mm	0	6% (?) *)

*) Only one sample, collected 1-2 m over Collingwood shale 5 km north of Pickering, Ontario, contains more rock fragments (shale) than minerals in these fine grades. This is a local till, consisting mainly of shale (100% shale in 4-8 mm grade); the boundary between minute shale fragments and individual clay minerals was not determined.

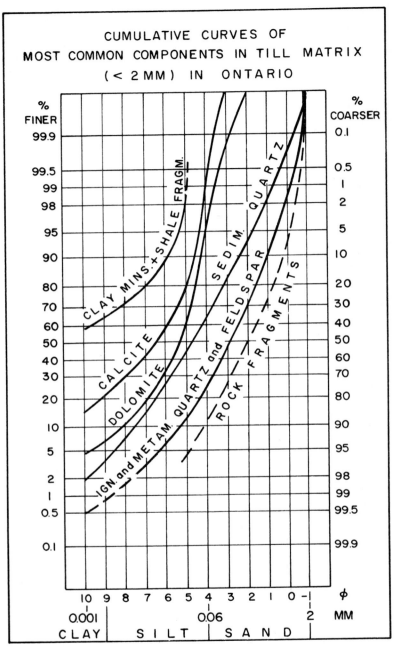

Fig. 7 Cumulative granulometric curves of most common components of till matrix (<2mm) in Ontario. The curves are averages of selected tills containing at least 6% of the corresponding component in the matrix: rock fragments from Nos. 1-30; igneous and metamorphic quartz — Nos 1-12; sedimentary quartz — Nos. 16-19 and 26-30; dolomite — Nos. 20-25, 27, and 29; calcite — Nos. 13-15, 18, 20, 21, 23, 25-27, 29, 30; clay minerals — Nos. 16-19.

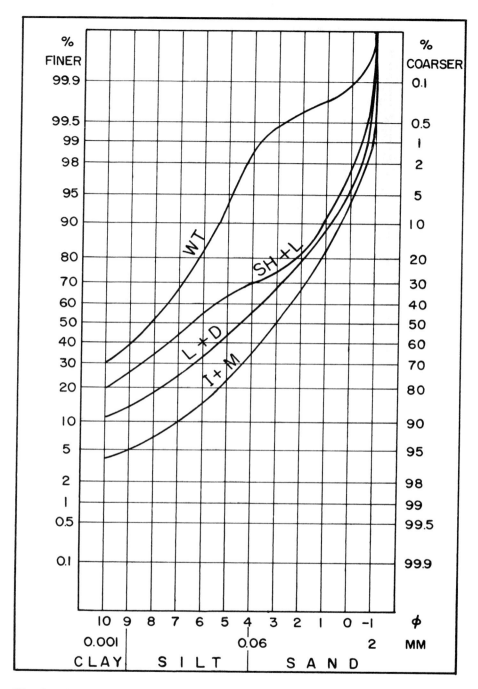

Fig. 8 Cumulative granulometric curves of the matrix of tills derived mainly from igneous and metamorphic rocks — I + M (Nos. 1-12), limestone and dolostone — L + D (Nos. 21-25), shale and limestone — SH + L (Nos. 26-28). The uppermost curve WT is an average of five waterlaid tills from the Port Stanley Drift in sourthwestern Ontario.

79

If shale is the only constituent, it may produce a silty clay till, if most shale fragments are soft and therefore crushed rapidly to their constituent minerals. In Ontario such shales are rare, as even the relatively soft shale formations e.g. Hamilton Shale, contain calcareous interbeds. Thus in Figure 8 the (SH+L) curve is an average for three tills (Nos. 26-28) collected from a Devonian shale area, but two of them (Nos. 26 and 27) were collected over interbedded shale and limestone of Hamilton Formation. Main constituent of the matrix of these three tills is rock fragments (an average of about 40%), mostly shale and limestone. Clay minerals, micas and tiny shale fragments of fine silt size are about 28%, carbonate minerals and silt-size quartz of sedimentary origin about 15% each, and minerals of igneous and metamorphic origin 3%. Abundance of silt-size sedimentary minerals combined with predominantly sand-size rock fragments has produced a sandy silt-till matrix, overshadowing the clay-and fine silt-size clay minerals.

Though knowledge of the mineralogical composition gives an indication of the type of texture of till matrix, the rock fragments which are particularly abundant in tills derived mainly from sedimentary fine-textured rocks, will generally cause the till to be coarser textured than would have been predicted by the mineral composition alone.

Incorporated stratified drift. Specific concentrations of minerals unrelated to the rules governing the glacial activities, may be found in tills which contain large amounts of incorporated sorted drift material, for instance the Port Stanley Till of south-western Ontario is either clayey or silty depending upon incorporation of lacustrine clays or silts, while the Wentworth Till (Karrow, 1963) is very sandy, particularly where outwash or kame sands and gravels have been incorporated. Waterlaid tills, formed most probably underneath an ice shelf, are usually very silty in the Port Stanley Drift of southwestern Ontario (Fig. 8: WT), containing large proportions of water-deposited silt-size particles.

CONCLUSIONS

Texture of till depends upon two major lithologic components: rock fragments and their constituent minerals. Most rock fragments occur in the coarse particle size classes: tills derived from igneous and metamorphic rocks contain them in particle sizes coarser than 1 mm, but tills derived mainly from sedimentary rocks - in the particle sizes coarser than 0.25 mm. Most textural analyses of tills in North America are done on till matrix, finer than 2 mm. Matrix of tills derived from igneous and metamorphic rocks depends mainly upon the terminal grades of feldspars and quartz, and therefore is coarse textured, with sand and coarse silt as main particle sizes. Matrix of tills derived mainly from sedimentary rocks is texturally more varied. Texture of these tills depends not only upon terminal grades of sedimentary minerals which are mainly of silt and clay sizes, but also upon the abundantly present fragments of medium hard sedimentary rocks, mainly of sand size. Besides, in southern Ontario, some tills contain igneous and metamorphic minerals up to one-third of the weight of their matrix. Texture of those tills which contain large proportion of incorporated stratified drift, depends mainly upon the dominant particle sizes of the incorporated material.

ACKNOWLEDGMENTS

The authors are grateful to the National Research Council of Canada for supporting their research by grant A4215; and to Mrs. R. Ringsman for drawing the illustrations.

REFERENCES

Dreimanis, A., 1969, Selection of genetically significant parameters for investigation of tills: Zeszyty Naukowe U.A.M., Geografia 8, Poznan, p. 15-29.

Dreimanis, A., and Vagners, U.J., 1969, Lithologic relationship of till to bedrock: in Wright, H.E. Jr. (ed.) Quaternary geology and climate, Nat. Acad. Sci., Publ. 1701, p. 93-98.

Dreimanis, A., and Vagners, U.J., 1971, Bimodal distribution of rock and mineral fragments in basal till: in Goldthwaite, R.P. (ed.): Till, A Symposium, Columbus, pp. 237-250.

Elson, J.A., 1961, The geology of tills: Proceed 14th Can. Soil Mich. Confer., Nat. Res. Counc. Can., Assoc. Com. Soil and Snow Mech., Techn. Mem. No. 69, p. 5-36.

Flint, R.F., 1971, Glacial and Quaternary geology, John Wiley and Sons, New York, 892 p.

Karrow, P.F., 1963, Pleistocene geology of the Hamilton-Galt area: Ont. Dept. Mines Geol. Rept. No. 16, 68 p.

Krumbein, W.C., 1934, Size frequency distributions of sediments: J. Sed. Petrology, v. 4, p. 65-77.

Lo, K.Y., and Roy, M., 1969, Rock breakage as related to some engineering and geologic processes, with discussion by Dreimanis, A.: Proceed. 22nd Can. Soil Mech. Confer., Queen's University, Dept. Civil Engin. Res. Rept. No. 67, p. 61-93.

Thomas, R.L. 1969. A note on the relationship of grain size, clay content, quartz and organic carbon in some Lake Erie and Lake Ontario sediments. Jour. Sed. Petrology, v. 39, p. 803-809.

Vagners, U.J., 1966, Lithologic relationship of till to carbonate bedrock in southern Ontario: unpublished M.Sc. thesis, Univ. Western Ontario, London, Ontario, 154 p.

Vagners, U.J., 1969, Mineral distribution in tills, south-central Ontario: unpublished Ph.D. thesis, University of Western Ontario, London, Ontario, 277 p.

Wentworth, C.K., 1922, A scale of grade and class terms for clastic sediments: J. Geol., v. 30, p. 377-392.

CLAY MINERALOGY OF GLACIAL CLAYS IN EASTERN DURHAM, ENGLAND

Peter Beaumont

INTRODUCTION

Since the middle of the nineteenth century the north-east coast of England has been correctly regarded as an area of conflict and intermingling of Quaternary ice-sheets derived from sources to the east, west and north (Beaumont 1968a) (Fig. 1). As a result of this history a complex sequence of glacial deposits mantles the area at the present day, varying in thickness from zero feet to more than 300 feet. This drift is thickest in the lowland zone and is often absent from areas above 500 feet on both the Permian escarpment and in the Pennine foothills. Five major glacial clays have so far been identified (Smith and Francis 1967; Beaumont 1968b). The oldest is a small exposure of Scandinavian Drift found only at Warren House Gill on the Durham coast (Trechmann 1915). The majority of the far travelled rocks in this deposit are thought to have originated in southern Norway. Overlying the Scandinavian Drift and forming the most widespread glacial clay in eastern Durham is the Lower Till. This clay usually rests on bedrock and forms a particularly persistent deposit, although it is rarely more than 30 feet in thickness. Throughout its outcrop it contains many pebbles of Carboniferous sandstone and limestone, together with material derived from the Lake District and Southern Uplands. Three divisions of the Lower Till sheet have been recognised dependent on the complexity of the glacial sequence. In the Wear Lowlands, the Lower Till generally forms the basal glacial deposit and is overlain by sands, laminated clays and stony clays. On the Magnesian Limestone Plateau to the east, the Lower Till sheet is the only widespread glacial deposit found at the present day. Finally, in the coastal area the Lower Till once more forms the basal member of a tripartite sequence of stony clays, sands and stony clays (Beaumont 1971).

In much of the coastal area the surface deposit is a stony clay known as the Upper Till. This deposit varies in thickness from 5 to 40 feet and contains rocks of Cheviot origin, together with Lake District and Southern Upland material. The Upper Till is almost always found overlying sands and gravels, with little evidence of disturbance or incorporation of these deposits. Forming the surface

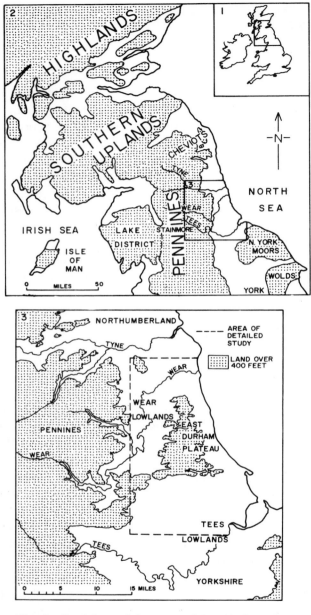

Fig. 1 Position of the area of detailed study.

deposit of much of the Tees Lowlands is the Upper Tees Clay. This deposit contains very few included stones and generally overlies laminated clays, sands and gravels and the Upper Till. In the Wear Lowlands a similar surface clay is found, which is known as the Upper Wear Clay and which shows considerable lateral variation from an almost stoneless clay, to a clay with a marked concentration of angular sandstone and coal fragments. It is usually only a few feet in thickness.

Of these clays the Scandinavian Drift, the Lower Till and the Upper Till are generally regarded as the basal deposits of ice sheets originating to the east, west and north respectively, of the eastern Durham area. The exact origin of the Upper Tees Clay and the Upper Wear Clay is somewhat more problematical as both these clays show considerable evidence of flowage phenomena (Beaumont 1970). Their glacial origin is, however, undisputed.

The ages of the three glacial tills - the Scandinavian Drift, the Lower Till and the Upper Till - are not definitely known owing to the lack of datable organic deposits within the region. A raft of Ipswichian peat has been discovered within the Upper Till of eastern Durham, and this enables the Upper Till to be dated as the product of the Devensian (Weichselian) glacial period (Beaumont, Turner and Ward 1969). Unfortunately the stratigraphical relationships at the site where the peat was discovered are such that it is not possible to infer the age of either the lower Till or the Scandinavian Drift. Recent carbon 14 dating of organic material in the Dimlington Silts of Holderness, which lie beneath the assumed lateral equivalents of the Upper and Lower Till sheets of County Durham (Hessle and Drab Tills of Holderness), establish an age of this material of c.18,300 years B.P. (Penny, Coope and Catt 1969). Both the Upper and Lower Till sheets of County Durham would, therefore, appear to have been deposited late in the Devensian glaciation probably between 25,000 to 15,000 years B.P. The Holderness equivalent of the Scandinavian Drift of eastern Durham, the Basement Till, is considered to be of Wolstonian age (Saale), but the possibility of an early Devensian age for the Durham deposit has not been conclusively disproved.

The clay content of the less than 2 millimetre fraction of the glacial clays of eastern Durham varies in the samples analysed from 13 to 43 percent, with a modal category value of between 25 to 30 percent. The Lower Till tends to be the most variable deposit in composition with clay contents varying from 13 percent to 32.5 percent, while the Upper Tees Clay reveals the highest clay contents with more than three-quarters of the samples possessing clay totals of greater than 35 percent.

The aims of the clay mineral analysis are as follows. First to discover what minerals are present within the glacial clays and in what proportions. Secondly, to evaluate whether the clay mineral suites of the different deposits can be used as distinguishing criteria in the identification of the glacial clays. Thirdly, to attempt to gain some ideas concerning the genesis of the clay mineral suites. Finally to assess the post-depositional effects of weathering on the mineral assemblages.

Method of analysis

In this study samples which were obtained from the five main clay groups are subjected to X-ray diffraction analysis and differential thermal analysis. Two kilogram samples of the glacial clays were collected in the field from cleared natural or man-made exposures. Whenever possible samples were collected at depths of six feet or more beneath the surface, and no samples were obtained at depths of less than four feet, except for specific weathering studies. In the laboratory material of less than 1.4 microns equivalent particle size diameter was separated by sedimentation techniques and prepared for X-ray diffraction and differential thermal analysis (Mackenzie 1955). With this method clay dispersion was facilitated by the addition of concentrated ammonium hydroxide to a mixture of 50 grams of the sample and distilled water. After a thorough stirring the mixture was transferred to a 1000 millilitre measuring cylinder, shaken end over end, and then allowed to stand. After 16 hours the top 10 cms was siphoned off. Distilled water was added to the cylinder and the procedure repeated until c.600 mls of the suspension was collected. This suspension was reduced in volume in a beaker on a water bath. When the mixture was reduced to a slurry a portion of the suspension was removed and bottled for the preparation of oriented samples for X-ray diffraction analysis. The remaining mixture was allowed to dry and then removed from the beaker. This was lightly broken up in an agate mortar and pestle and passed through a British Standard No. 72 mesh sieve prior to storage.

X-ray diffraction analysis

In the initial stages two types of mount for each sample were prepared for analysis. Oriented mounts were made by placing a small portion of the clay slurry on a glass slide and allowing it to dry. Unoriented mounts, or cavity mounts, were prepared by packing the dried clay sample into an aluminium holder. These samples were analysed by X-ray diffraction using a Philips diffractometer and CuKα radiation. Both oriented and unoriented samples were scanned between 2 to 62 degrees 2θ at 1 degree 2θ per minute. The diffraction angles 2θ are converted to d spacings in Angstrom units by means of tables or graphs and then compared with the d spacings of standard minerals recorded by the American Society for Testing Materials (A.S.T.M.). A number of the commonly occurring clay minerals can usually be identified at this stage.

The presence of illite/muscovite mica is revealed by a large broad peak between 9 to 10 Angstrom units (A) and a smaller peak at 4.45 A. Kaolinite shows peaks at 7.16 A, 4.45 A, 4.18 A and 3.57 A, and quartz at 4.26 A and 3.33 A. Chlorite exhibits a 14 A peak and other peaks at 7.07 A, 4.72 A and 3.54 A. Minerals in the montmorillonite group are usually identified by a peak between 12 to 15 A. Mixed layer minerals, particularly randomly interstratified clays, are usually more difficult to identify with any precision owing to the possible complexity of the structure. Mixed layering of mica is generally indicated by a series of reflections on the high A side of the 10 A d spacing. The resultant spacing is broad and has several peaks making up the hump or shoulder (Carroll 1970). Mixed layering of montmorillonite and chlorite is indicated by reflections on the lower A side of the 14 A d spacing. Mixed layer or intergrown illite-montmorillonite will have peaks falling intermediate between those of illite 10 A and of montmorillonite 12 to 15 A. The exact position of these peaks depends on the relative amounts of the two components (Weaver 1958).

Other treatments of the sample to aid mineral identification are glycolation and heating. Both these treatments cause changes in certain minerals which are measurable on the diffractogram. Glycolation is carried out by placing a mounted sample in a desicator containing a small quantity of ethylene glycol (Brunton 1955). To aid expansion of the minerals the desicator can be placed in an oven at 60 degrees C. for an hour or longer. When cool the sample can be analysed and a diffractogram obtained. The identification of some minerals can be facilitated by heating the sample to different temperatures in a thermostatically controlled muffle furnace (Carroll 1970 Table 10). A diffractogram is then made from the sample for subsequent analysis.

In this study the identification of quartz, illite/muscovite mica and kaolinite proved relatively straightforward with the initial oriented and unoriented diffractograms. The identification of chlorite, in the presence of kaolinite; of montmorillonite in small quantities; and of mixed layer minerals, proved much more difficult. The distinction between kaolinite and chlorite, when no 14 A peak of chlorite is visible, can sometimes be made following heat treatment of a sample to 550 to 600 degrees C. This makes kaolinite become amorphous, and hence reveal no diffraction pattern, while at the same time the intensity of the 14 A peak of chlorite usually increases. This technique does not, however, always produce satisfactory results. Therefore, whenever the presence of chlorite is suspected in a sample a slow scanning technique at ¼degree 2 θ per minute between 22 to 28 degrees 2 θ is employed in an attempt to separate the 3.54 A peak of chlorite from the larger 3.57 A peak of kaolinite (Biscaye 1964). With this method it is often possible in the samples analysed to distinguish the two peaks, and so show the occurrence of chlorite in association with kaolinite. A number of samples, although not revealing a distinct 3.54 A chlorite peak do show a broad shoulder on the much larger 3.57 A kaolinite peak suggesting that chlorite is possibly present though in very small quantities. Glycolation of the sample causes some clay minerals to expand and so provides a further aid to identification. Kaolinite, illite and chlorite are all unaffected following treatment by ethylene glycol. Following glycolation, montmorillonite shows a peak movement from between 12 to 15 A, to 17 A, while mixed layer minerals, with untreated peaks between 10 to 15 A react slightly differently depending on what minerals are present within the layers. With mixed layer chlorite-montmorillonite the glycolated sample tends to reveal a broad 14 to 17 A peak. Mixed layer illite-montmorillonite is more difficult to identify with certainty especially if montmorillonite layers predominate. If, following glycolation, the peak does not shift completely to 17 A it can be assumed that some non-montmorillonite layers, illite or chlorite are present (Weaver 1958).

Clay mineral identification

1. Lower Till (Figure 2)

The mineralogy of the Lower Till reveals pronounced similarities in its three divisions, with all of the samples characterised by the presence of illite/muscovite mica, kaolinite and colloidal quartz. A number of the samples show a peak at 14.2 A which could be produced by chlorite, montmorillonite or mixed layer minerals. In almost all cases, even when no definite 14 A peak is seen, slow scanning between 22 to 28 degrees 2 θ reveals the presence of a separate chlorite peak at 3.54 A or at least a hump on the shoulder of the 3.57 A kaolinite peak. Heat treatment of samples which do not reveal a 14 A peak

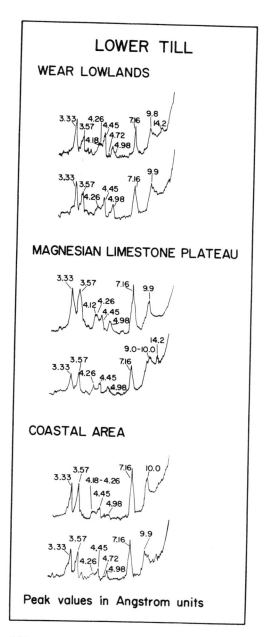

Fig. 2 X-ray diffractograms of oriented samples from the Lower Till.

sometimes causes one to appear. As a result of this analysis it is considered that chlorite is probably present in all the samples although almost always in very small quantities. The nature of the 9 to 10 A peak varies considerably. In some samples a very pronounced peak occurs suggesting the possible presence of fine grained muscovite (White 1962). Many of the samples on the other hand exhibit a broad shoulder to the 10 A peak or else a very diffuse pattern in this region, suggesting the presence of mixed layer minerals. Glycolation of these samples does not, however, produce any clear new peaks at higher A values which could be differentiated from the general background of the diffractogram. Such evidence would seem to point to the fact that the mixed layer minerals are of only minor importance.

2. Upper Till (Figure 3)

The Upper Till shows a very similar mineralogy to that of the Lower Till, with illite/mica, kaolinite and quartz the dominant minerals present. In most of the samples both the 14.2 A peak and the 3.54 A peak of chlorite are better developed than in the Lower Till, indicative of the occurrence of chlorite in perhaps greater amounts. The 9.8 illite/mica peak in the Upper Till also tends consistently to be sharper than in the Lower Till. This would suggest the lesser importance of mixed layer minerals or of higher concentrations of illite/mica. Neither of the above differences are, however, sufficient to allow any positive differentiation between the two deposits.

3. Upper Tees Clay (Figure 3)

Kaolinite and quartz are easily recognised constituents of this clay. A marked feature of the Upper Tees Clay is the presence of considerable quantities of mixed layer minerals (diffuse peaks at 9 to 14 A) and a lesser quantity of illite/mica than in the Upper and Lower Tills. Slow scanning at 22 to 28 degrees 2 θ shows the presence of chlorite and this is further collaborated by heat treatment. Once again glycolation does not produce diagnostic results, although a very diffuse peak between 14 to 16 A could have indicated the presence of a mixed layer chlorite-montmorillonite mineral. In such samples it would seem likely that a number of different mixed layer types might be present. Glycolation does produce a clear illite/mica peak at 9.9 A.

4. Upper Wear Clay (Figure 3)

The mineralogy of the Upper Wear Clay is similar to that of the Lower Till, with illite/mica, mixed layer minerals, kaolinite, quartz and chlorite present in all samples. One difference occurs,

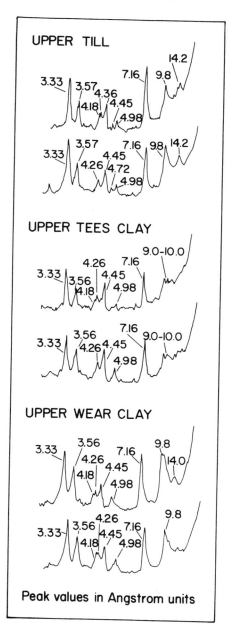

Fig. 3 X-ray diffractograms of oriented samples from the Upper Till, the Upper Tees Clay and the Upper Wear Clay.

however, in that a number of the samples on glycolation reveal a new, small but sharp peak at 17.8 A. This peak would seem to be at too high an A value to be explained as being due to the occurrence of mixed layer minerals, and would therefore seem to indicate the presence of a pure montmorillonite.

5. Scandinavian Drift (Figure 4)

The Scandinavian Drift is the only glacial clay within eastern Durham which shows any variation from the mineral sequence as set out above for the other clay deposits. Even in this case, the differences are only of a minor nature. With this deposit the diffractogram is dominated by an extremely well marked quartz peak to a degree not observed in the other Durham glacial clays. Illite/mica, together with mixed layer minerals are present, as indicated by a series of diffuse peaks between 10 to 14 A. Once again chlorite is found but only in small quantities. The 7.16 A peak is small and this fact coupled with the weakness of both the 4.1 A and the 3.57 A peaks suggests that kaolinite occurs in smaller proportions than in the other Durham glacial deposits. Of special interest with this sample are two peaks at 8.4 A and 3.03 A which are not usually observed in the Durham clays. Both of these peaks represent rock minerals rather than clay minerals. The 8.4 A peak is produced by an amphibole and, to some extent, confirms the supposition that this deposit has at least been partially derived from igneous or metamorphic material. The 3.03 A peak is due to the presence of calcite, probably derived from the grinding down of chalk material which is found in this deposit.

6. Svartisen Glacier (Figure 4)

For comparitive purposes a till deposit from the present day Svartisen Glacier, Norway was analysed. In marked contrast to the glacial clays of eastern Durham this sample is almost exclusively composed of rock minerals rather than clay minerals. The Svartisen sample is essentially "rock flour" made up of muscovite mica (peak at 9.8 A), feldspar (peak at 3.23 A) and quartz (peak at 3.33 A). A small peak at 8.4 A indicates the presence of an amphibole, while the 3.03 A peak suggests calcite. Chlorite occurs in small quantities. No other mineral appears to be present in any quantity.

From the above descriptions it is seen that the differences in clay mineralogy between the glacial clays of eastern Durham as revealed by X-ray diffraction analysis are only minor and cannot be used for identification purposes with any degree of certainty. In all the glacial clays illite and mixed layer minerals, kaolinite,

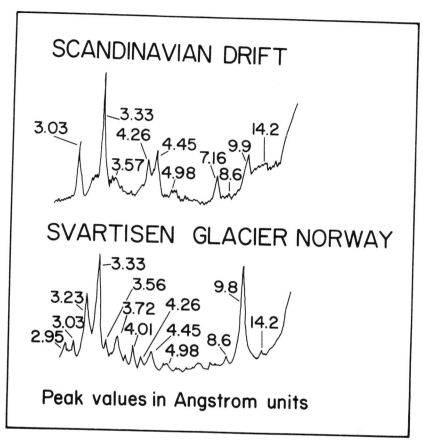

Fig. 4 X-ray diffractograms of oriented samples from the Scandinavian Drift and the Svartisen Glacier, Norway.

quartz and chlorite are found. Only montmorillonite appears to be of restricted occurrence (Table 1).

Differential Thermal Analysis

To complement the X-ray diffraction analysis of the glacial clays, powder samples (less than 1.4 microns) were subjected to differential thermal analysis. A machine designed and built by Bolton's of Edinburgh was employed. With this system a sample of the material to be analysed is placed in one hole of the sample holder and an inert material in the other. Thermocouples are attached to the centres of both these holders and the holders together with the thermocouples are placed in a furnace controlled to produce a uniform temperature rise of 10 degrees C. per minute. To avoid oxidation of the sample nitrogen is passed through the furnace while the experiment is in progress. All reactions during which heat is lost or gained by the sample (changes in water content, recrystallisation, changes in structure etc.) are detected by the thermocouples and recorded as a curve on a chart. In these curves exothermic reactions are recorded by deflections in an upward direction and endothermic reactions by a downwards deflection. Differential thermal analysis curves are analysed by comparison with published mineral curves or those of standard minerals (Mackenzie 1957, 1962).

Following differential analysis illite reveals an initial endothermic peak at 125 degrees C., which is generally small and a much larger endothermic reaction beginning at 450 degrees C. and peaking between 600 to 650 degrees C. Between 950 to 1000 degrees C. a strong exothermic reaction occurs. Kaolinite exhibits a large endothermic peak beginning at about 400 degrees C. and reaching a maximum between 600 to 650 degrees C. A pronounced exothermic reaction is noted at 900 to 1000 degrees C. Quartz registers a single and sharp endothermic reaction at 573 degrees C. Chlorite is characterised by a pronounced endothermic reaction between 500 to 700 degrees C. followed by a second endothermic peak at 800 degrees C. Immediately following this is a pronounced exothermic reaction between 830 to 900 degrees C.

These results illustrate the difficulty of using D.T.A. for the identification of clay minerals in mixtures owing to the fact that similar reaction patterns are shown by the common minerals. For example, illite, kaolinite and quartz all produce endothermic reactions between 570 to 650 degrees, while both illite and kaolinite reveal exothermic reactions close to 950 degrees C. Indeed only the trace of chlorite is sufficiently different to suggest the

TABLE 1

OCCURRENCE OF CLAY SIZED MINERALS

	LOWER TILL	UPPER TILL	UPPER TEES CLAY	UPPER WEAR CLAY	SCANDI- NAVIAN DRIFT	SVARTI- SEN GLACIER
Illite/mica	1	1	1-2	1	1-2	1
Kaolinite	1	1	1	1	2	4
Quartz	1	1	1	1	1	1
Chlorite	2-3	2	2-3	2-3	2-3	3-4
Mixed-layer clays	2-3	3	2	2-3	3	4
Montmorillonite	4	4	4	3-4	4	4
Amphibole	4	4	4	4	2-3	2-3
Calcite	4	4	4	4	1	2-3
Feldspar	4	4	4	4	4	1

1-Easily identified; 2-Average identification; 3-Marginal identification; 4-Not detectable
(This table cannot be taken as an indication of the relative importance of the minerals present)

possibility of positive identification using D.T.A. As it happens this is the one mineral that it consistently proved difficult to identify positively in many of the glacial clays. It was, therefore, hoped that D.T.A. would be able to prove conclusively the presence or absence of chlorite in samples.

Differential thermal analysis traces of the five main glacial clays of eastern Durham reveal very marked similarities (Figure 5). Every curve shows a small endothermic peak at 125 degrees C. (illite), a large endothermic peak which reaches a maximum close to 630°C. (illite, kaolinite and quartz), and an exothermic peak at 960 degrees C. (illite and kaolinite). Most samples also reveal a small endothermic peak at 300 degrees C. of unknown origin. No sample shows evidence of any chlorite peak even though X-ray analysis has revealed its presence in a number of samples. This fact further strengthens the idea that when chlorite occurs within a sample it does so only in very small quantities.

Semi-quantitative estimation of the clay minerals

Quantative evaluation of clays is extremely difficult and usually only rough estimates of the proportions of the minerals present can be made (Weaver 1958, Brown 1961), even though many different methods have now been employed (Carroll 1970). In this study standard minerals obtained from the Macaulay Institute were mixed in varying proportions to compare with the mineralogy of the glacial clays. These mixtures were subjected to X-ray diffraction analysis and the diffractograms collated with those of the glacial clays. The greatest problem with this method is the fact that the crystallinity and even the composition of the standard mineral composing the mixture may be considerably different from similar minerals present in the glacial clays. Such differences would produce variations in peak intensities of unknown amounts. The following study was undertaken, therefore, only to obtain an approximate indication of the relative proportions of the minerals present.

A number of differing mixture compositions were prepared and analysed. One of these, a mixture of 25 percent quartz, 25 percent kaolinite, 25 percent illite and 25 percent chlorite is shown in Figure 6 to illustrate the relative crystallinity of the standard minerals. The peaks occurring in this sample are similar in position to those of the glacial clays, but the relative heights of the peaks show considerable differences. In particular illite appears to be greatly underrepresented and chlorite over represented. Of the mixtures analysed it was found that two were similar to many of the diffractograms of the glacial clays (Figures 2 and 3). These mixtures had compositions by weight as indicated in Table 2.

TABLE 2

COMPOSITION OF MIXTURES BY WEIGHT

	Mixture A (%)	Mixture B (%)
Illite	60	72.5
Kaolinite	20	20
Quartz	15	5
Chlorite	5	2.5

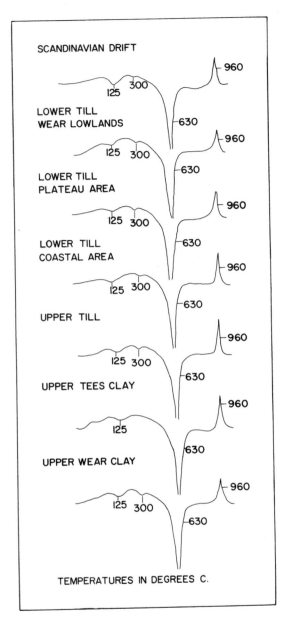

Fig. 5 Differential Thermal Analysis curves of samples from the glacial clays

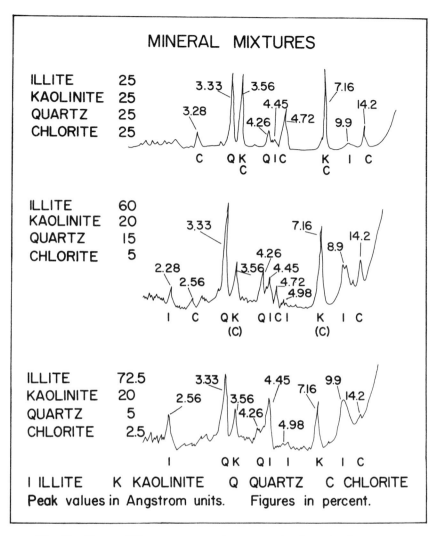

Fig. 6 X-ray diffractograms of prepared mineral mixtures.

It appears, therefore, that illite (together with mixed layer minerals) is probably the dominant clay mineral present in the glacial clays of eastern Durham. Actual quantities present in the different samples possibly vary from 40 to 70 percent, but it does seem likely that in most samples illite (and mixed layer clays) make up more than 50 percent of the total mineral assemblage. Kaolinite would seem to be the second most important mineral in the glacial clays. It shows less absolute variation than illite with the majority of the samples probably containing between 15 to 25 percent kaolinite. In a few samples greater percentages of kaolinite possibly do occur. Despite the height of the peaks produced by quartz this mineral occurs in quantities of less than 15 percent in almost every case. A figure of 5 to 10 percent might be a modal value of quartz in the field samples. What is of interest is the fact that quartz is detectable in every sample analysed even at this particularly fine grain size of less than 1.4 microns. If chlorite occurs in a sample it is present in quantities of less than 10 percent and possibly even less than 5 percent. In the standard mixtures, chlorite, when present in concentrations of less than 5 percent, is detectable only with great difficulty. On the other hand if the chlorite of the glacial clays is less crystalline than the standard mineral, it might not be detected in concentrations of less than 15 percent with any degree of certainty.

Origin of the clay minerals

The great similarity of the clay mineral content of the glacial clays of eastern Durham is one of the main features noted in this study. The question arises as to whether this mineralogy is explicable in terms of the environment of deposition (i.e. wetting of the clay minerals in the metamorphic environment at the base of the glacier) or whether all the clay minerals have been derived solely from the erosion of nearby rock types. It is also possible that the clay minerals may have been formed by the weathering of rock debris subsequent to the deposition of the glacial clays.

To test the second hypothesis, studies of the clay mineralogy of the local parent materials were made. By far the largest rock outcrop within eastern Durham is composed of Magnesian Limestone. X-ray analysis of this rock reveals that the only minerals present in any quantity are dolomite and calcite (Figure 7). Therefore, the contribution of this rock type to the mineralogy of the glacial clays is minimal. Analysis of the glacial clays also reveals that the calcite and dolomite are not usually broken down to fine grain sizes (less than 1.4 microns) as the result of ice

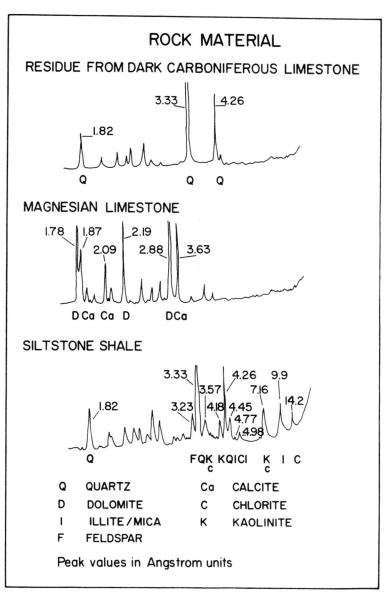

Fig. 7 X-ray diffractograms of local rock samples.

action. The only other limestones close by are the Lower Carboniferous limestones of the Pennines. These are composed mainly of calcite, but the dark colours of many of them suggests the presence of impurities. To discover what these impurities are pieces of such limestone were dissolved in weak hydrochloric acid and the insoluble residue subjected to X-ray analysis. The analysis proved that most of the insoluble residue within these limestones is quartz with minor amounts of illite and mixed layer minerals (Figure 7).

The key to the mineralogy of the glacial clays of eastern Durham is to be found in the Carboniferous Coal Measures which encircle eastern Durham to the north and west. The Coal Measures consist largely of sandstones, siltstones and shales and of these the siltstones and shales are the most likely contributors to the clay sized material of the glacial clays. A number of siltstone and shale samples were X-rayed and all proved to possess similar mineralogies. A representative sample trace is shown in Figure 7. In this sample all the clay minerals commonly found in the glacial clays of eastern Durham are seen. Owing to the fact that no size separation of the shale was feasible prior to analysis it is noted that quartz dominates the diffractogram. Illite is present in considerable quantities and owing to the extreme sharpness of the 9.9 A peak, muscovite is also thought to occur. Kaolinite is present, although in this particular sample not in very large quantities. Chlorite is also an important and easily recognisable constituent. The only mineral present in the shale sample which is not commonly found in the glacial clays is feldspar (peak at 3.23 A). This is probably owing to the fact that this mineral occurs in the size range of more than 1.4 microns in the shale sample.

To further test this hypothesis that the clay minerals of the glacial clays were derived directly from local parent material, it was decided to form an "artificial" glacial clay by mixing local parent materials and then to analyse the clay mineralogy of this sample. Unweathered rock specimens were collected from exposures of the Coal Measures within the Wear Lowlands. Nine subsamples of approximately equal size were selected - three of sandstone, three of siltstone and three of shale - from widely differing parts of the region. These rocks were crushed between metal plates until all the rock fragments passed through a ¾inch British Standard Sieve. The mixture was then placed in a metal container and shaken for 6 hours in an attempt to simulate the grinding and mutual attrition of material which is thought to take place in the basal layers of a glacier during transport. Following this treatment the clay sized material (less than 1.4 microns) was separated by sedimentation and subjected to X-ray diffraction analysis. The resulting diffractogram reveals the presence of illite (and mixed layer minerals), kaolinite, quartz and chlorite (Figure 8). The proportions of these minerals are very similar to those found in the glacial clays and in particular to one of the mixtures of the standard clay minerals.

Finally an assessment was made of the weathering effects on the clay minerals since the deposition of the glacial clays. For this study samples of the Lower Till were obtained at 1 foot intervals from vertical exposures. Clay sized material was separated out and subjected to X-ray diffraction analysis. Of the profiles studied, most of them exhibit little evidence of changes in the mineral patterns except within 2 feet or so of the ground surface. At Coxhoe Quarry the Lower Till of the plateau area is c. 8 feet thick and rests directly on Magnesian Limestone. The diffractograms at the different depths are almost identical with only the uppermost and lowermost samples showing any differences (Figure 9). The sample at 1 foot depth is unusual in possessing a well marked peak at 6.26 A, believed to be due to the iron mineral lepidocrocite. The presence of this mineral would suggest fairly intense weathering conditions at this level. The 7 foot sample shows a peak at 2.88 A, which indicates fine grained dolomite. Undoubtedly this originated from the Magnesian Limestone parent material on which the clay rests. All the diffractograms are characterised by the occurrence of illite/mica, kaolinite, quartz and almost certainly chlorite.

CONCLUSIONS

This work has shown that the clay mineralogy of the four major glacial sheets of eastern Durham is extremely similar. Only the Scandinavian Drift possesses a slightly different composition, but even in this case, almost all of the minerals present are also found in the other clays. Illite/mica (with mixed layer minerals) is the dominant clay mineral present and would seem to make up more than 50 percent of individual samples in many cases. Kaolinite is the second most important mineral followed in turn by quartz. Chlorite appears to be present in probably all samples, but always in small proportions. Montmorillonite is detectable in only a few samples of the Upper Wear Clay. Analysis of the mineralogical composition of local rock material and the "artificial" till reveals that all the minerals in the clay sized fraction of the four main locally formed glacial clays were derived almost exclusively from the sandstones, siltstones and shales of the Coal Measure Series. No evidence is available to suggest that the sub-glacial transport environment has had any effect on the clay mineral suites present in the glacial deposits. Finally it would seem that post depositional

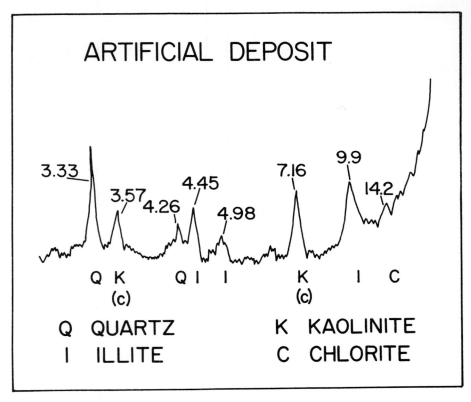

Fig. 8 X-ray diffractogram of oriented samples of artificial deposit.

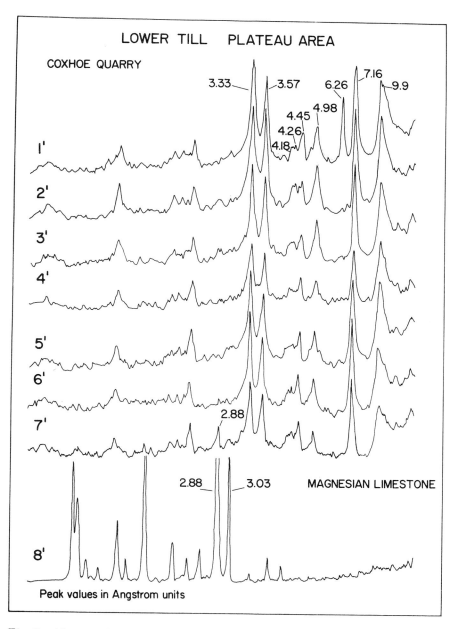

Fig. 9 X-ray diffractograms of oriented samples from the Lower Till at Coxhoe Quarry.

weathering of the glacial clays has only had a minor effect upon the clay sized material and this only in the zone closest to the ground surface.

REFERENCES

Beaumont, P., 1968(a): *The Glacial Deposits of Eastern Durham.* Unpublished Ph.D. Thesis, University of Durham, England.

Beaumont, P., 1968(b): *A History of Glacial Research in Northern England.* Department of Geography, University of Durham, Occasional Paper Series No. 9, 21 pp.

Beaumont, P., Turner, J., and Ward, P.F., 1969: An Ipswichian Peat Raft in Glacial Till at Hulton Henry, Co. Durham. *New Phytologist,* v. 68, pp. 797-805.

Beaumont, P., 1970: *Geomorphology.* In *Durham County and City with Teeside,* Dewdney, J.C. (ed.) British Association for the Advancement of Science, Durham, pp 26-45.

Beaumont, P., (1971 - In press): Stone orientation and stone count data from the Lower Till Sheet, Eastern Durham, England. Proceedings of the Yorkshire Geological Society.

Biscaye, P.E., 1964: Distinction between kaolinite and chlorite in recent sediments by X-ray diffraction. *American Mineralogist,* v. 49, pp. 1281-1289.

Brown, G. (ed.), 1961: *The X-ray identification and crystal structures of clay minerals.* Mineralogical Society (Clay Minerals Group), London, 544 pp.

Brunton, G., 1955: Vapor pressure glycolation of oriented clay minerals. *American Mineralogist,* v. 40, no. 1-2, pp. 124-126.

Carroll, D., 1970: *Clay Minerals: A Guide to Their X-ray Identification.* The Geological Society of America, Special Paper 126, 80 pp.

Mackenzie, R.C., 1955: Methods for separation of soil clays in use at the Macaulay Institute for soil research. *Clay Minerals Bulletin,* v. 3, no. 15, pp. 4-6.

Mackenzie, R.C. (ed.), 1957: *The Differential Thermal Investigation of Clays.* Mineralogical Society (Clay Minerals Group), London, 456 pp.

Mackenzie, R.C., 1962: *Scifax Differential Thermal Analysis data index.* Cleaver-Hume Press Ltd., London.

Penny, L.F., Coope, G.R., and Catt, J.A., 1969: Age and insect fauna of the Dimlington Silts, East Yorkshire. *Nature,* v. 224, no. 5214, pp. 65-67.

Smith, D.B. and Francis, E.A., 1967: *Geology of the country between Durham and West Hartlepool.* Memoirs of the Geological Survey of Great Britain, New Series Sheet 27, H.M.S.O., London, 354 pp.

Trechmann, C.T., 1915: The Scandinavian Drift of the Durham Coast and the general glaciology of South-east Durham. *Quarterly Journal of the Geological Society,* v. 71, pp. 53-82.

Weaver, C.E., 1958: Geologic Interpretation of Argillaceous Sediments; Part I - Origin and Significance of Clay Minerals in Sedimentary Rocks. *Bulletin American Association of Petroleum Geologists,* v. 42, pp. 254-271.

White J.L., 1962: X-ray diffraction studies on weathering of muscovite. *Soil Science,* v. 93, pp. 16-21.

COMPUTER APPLICATIONS IN THE ANALYSIS OF HEAVY MINERAL DATA FROM TILLS

Hugh Gwyn and P.G. Sutterlin

INTRODUCTION

With the advent of computer technology, the collection, storage, retrieval and analysis of data of many kinds by computer has become practical. Many workers, however, still question the advisability of using computer techniques for these purposes due to the costs involved, particularily when applied to a relatively small set of data. However, before consideration can be given to building large computer processible files of data, test files must be built in an attempt to assess types of information that should be recorded for a particular study and the utility of this data whether in computer processable form or not. The purpose of this case history is, then, to assess the applicability of computer techniques using a relatively small set of data of heavy minerals in tills.

A suite of 132 till samples was collected for this study. 115 of these samples were collected from tills on the periphery of the Canadian Shield in the Great Lakes region (Fig. 1). The purpose was to attempt to define the heavy mineral content of tills at the point where the glacier, which deposited and transported the tills, began over-riding Paleozoic rocks in Southern Ontario. It was reasoned that if significant differences in the heavy mineral content could be demonstrated, the imprint of this heavy mineral content might be used as an indicator of provenance of tills south of the Shield periphery.

The initial consideration was of course, what data in addition to the heavy mineral data should be recorded for each sample. A list was compiled (TABLE 1) of those items of data which were considered significant and necessary in the description of each sample. The first group of data items pertain to the identification and location of the sample. The second group of data items concern the geological setting and nature of the contiguous bed rock lithology. The third group of data items are concerned with the heavy mineral content of each sample.

TABLE 1 also serves as the Data Specifications List (or file description) to the SAFRAS system. The number to the left of the heading for each particular group of data items (e.g. *0101) denotes the data category as the first two digits and the number of times this data category occurs for each sample in the file as the

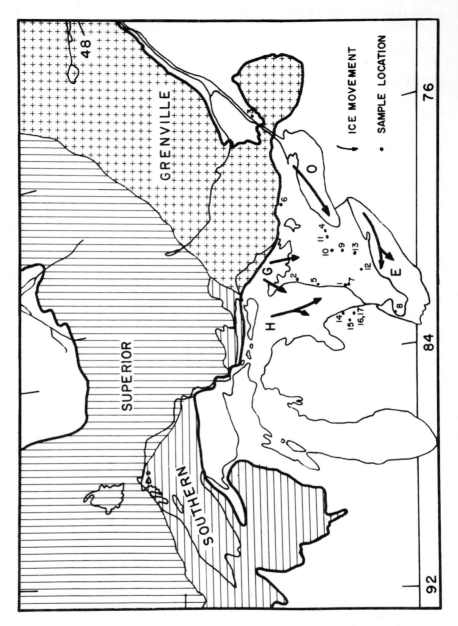

Fig. 1 Map of Precambrian structural provinces showing till sample locations on Paleozoic terrane, and direction of Lake Wisconsin ice movement (E — Erie lobe, G — Georgian Bay lobe, H — Huron lobe, O — Ontario lobe).

TABLE 1

LIST OF INFORMATION STORED FOR EACH TILL SAMPLE (DATA SPECIFICATIONS LIST)

*0101   SAMPLE-LOCATION

ID-NUMBER	5	N
FIELD-NUMBER	12	X
LAT-DEGREES	2	N
LAT-MINUTES	2	N
LAT-SECONDS	2	N
LAT-DIRECTION	1	A
LONG-DEGREES	3	N
LONG-MINUTES	2	N
LONG-SECONDS	2	N
LONG-DIRECTION	1	A
SAMPLE-TYPE	10	A
COMMENTS	30	X

*0204   BEDROCK-GEOLOGY

GEOLOGIC-AGE	4	A
LITHOLOGY	8	A

TABLE 1 (cont.)

*0303 HEAVY-MINERALS

Code	Description			
UPPER-GRAIN-SIZE		1	03	N
LOWER-GRAIN-SIZE		1	03	N
WTHV	(weight heavy minerals)	2	02	N
MGNT	(magnetic minerals)	2	02	N
TOTAL-GRAINS		3		N
HBLD	(hornblende)	3		N
BSHB	(basaltic hornblende)	3		N
TRML	(tremolite-actinolite)	3		N
RBCK	(riebeckite)	3		N
OPRX	(orthopyroxene)	3		N
CPRX	(clinopyroxene)	3		N
RGRN	(red garnet)	3		N
PGRN	(purple garnet)	3		N
ZOIS	(zoisite)	3		N
EPDT	(epidote)	3		N
RUTL	(rutile)	3		N
SPHN	(sphene)	3		N
ZRCN	(zircon)	3		N
OPAQ	(nonmagnetic opaque minerals)	3		N
MICA	(mica)	3		N
TMLN	(tourmaline)	3		N

second two digits. Therefore, 0101 would indicate that the items which follow are in the first group of data items and that this group of data occurs only once for each sample. In contrast, the second group of data items can occur up to four times for each sample and the third group of data can occur up to three times for each sample. Using SAFRAS terminology, each group of data items is termed "an element set". SAFRAS will allow 1 up to ninety-nine different element sets and will also permit the occurrence of each element set up to ninety-nine times. The numbers in the three right hand columns, reading from left to right denote the number of characters reserved in the file for that item of data, the number of decimal points (if any) in the data and the nature of the data (where N signifies numeric data, A signifies data comprising alphabetic characters only, and X denotes data or information which contains both alphabetic and/or numeric data).

DATA REDUCTION

Using this Data Specification List (file description), a document on which the data is recorded can be prepared either automatically by computer or manually (Table 2). On this document are printed the names of the data elements on alternate lines deliniated by slash marks. Under each of the item names is a blank field also deliniated by slash marks in which the data elements are entered. The only restriction in recording elements is that the actual data be limited to the area between slash marks. Necessary right or left justification and additional formating is done automatically by the SAFRAS system. This form, which in SAFRAS terminology is termed the "Source Document", can be used to directly reduce the data by key punching onto 80 column punch cards. The data are punched as a linear string beginning in column 1 of Card 1 and continuing, assuming that this card is infinite in length. The data items are not referenced to any specific columns on any specific card.

Experience has shown that data recording in this manner is not only rapid but accurate, and key punching becomes efficient and relatively error free in comparison to more conventional fixed-format data entry methods.

DATA FILE GENERATION

The data in element sets 1 and 2 (Table 1) were derived from field notes and maps, both for location and geology. The purpose of repeating the Type 2 element set four times, is in order that the bed rock types at varying distances up ice from the sample location

TABLE 2

DATA SOURCE DOCUMENT

Code	/ID Number (5N)	/Field Number (10X)	/Lat Deg (2N)	/Lat Min (2)	/Lat Sec (2N)/
*0101	/	/	/	/	/

Lat Dir (1A)	/Long Deg (3N)	/Long Min (2N)	/Long Sec (2N)	/Long Dir (1A)/
/	/	/	/	/

Sample Type (10A)/	Comment (30X) /

Code	/Geologic Age (4X)	/Lithology (8X)/	Code	/Geologic Age (4X)	/Lithology (8X)/
*0201	/	/	*0202	/	/

Code	/Geologic Age (4X)	/Lithology (8X)/	Code	/Geologic Age (4X)	/Lithology (8X)/
*0203	/	/	*0204	/	/

TABLE 2 (cont.)

Code	UGS (1.3N)	LGS (1.3N)	WTHV (2.2N)	MGNT (2.2N)	Total (3N)	HBLD (3N)
*0301						

BSHB (3N)	TRML (3N)	RBCK (3N)	OPRX (3N)	CPRX (3N)	RGRN (3N)	PGRN (3N)

ZOIS (3N)	EPDT (3N)	RUTL (3N)	SPHN (3N)	ZRCN (3N)	OPAQ (3N)	MICA (3N)

TMLN (3N)

Note: See Table 1 for list of abbreviations.

could be assessed to determine their relative influences on the heavy mineral content. The purpose of repeating the type 3 element set three times is that heavy mineral point counts were performed on three separate grain sizes, fine sand, very fine sand, and coarse silt. The choice of these three grain sizes for heavy mineral analysis is based on the work of Vagners (1969).

The heavy mineral data was acquired by performing counts using a petrographic microscope with a mechanical stage attachment. Between 450 and 550 individual grains were identified and counted for each grain size in each sample. The mineral species and groups identified, coded as four letter mnemonic abbreviations, are listed in Table 1. A total of 86 individual items of information were recorded for each sample. The data file then consists of nearly 12,000 separate items of information for the 132 samples.

The data punched on cards together with the file description and the SAFRAS system programs were then used to generate a formatted computer-processible file from which selective retrievals of various kinds were performed in order to:

1. Visually inspect various groups of samples and/or data in order to draw some conclusions with regard to their distributions, and

2. to extract various combinations of numerical data for further analysis using mathematical techniques.

Without going into detail on the actual SAFRAS system itself, suffice to say that once the file is generated any item of data in the file can be used as a criterion for retrieval in a conditional retrieval request. In addition, any or all items of information can be printed, written on tape or punched on cards for further analysis. The flexibility of the system allows the researcher to add data, edit the file, and provides the ability to augment the content of a file or merge files without the necessity of any re-programming.

DATA RETRIEVAL

The first retrievals from the file can be characterized as generalized qualitative retrievals. They are basically of two types based on:

1. the geology or lithilogic associations of the till samples and,

2. their association with rocks of various geological ages.

116

Table 3 illustrates the request submitted for one of these retrievals and a partial listing of the output. Listings such as the one illustrated were inspected visually to detect any empirical correlations or associations in the data. For example, it becomes obvious that when associated rock types are "basic igneous", there is a relatively high concentration of magnetic minerals and clinopyroxene in the tills. Other retrievals of this same general character pointed out the following:

1. There is an association of the heavy mineral content of the till samples and the grade of metamorphism of the local bed rock.

2. With respect to the content of magentic minerals, clinopyroxene, and hornblende, the boundaries between the Superior and Southern, as well as the Southern and Grenville Provinces, could be related to the locations of the till samples (Fig. 1).

3. Using the Grenville Province as a retrieval criterion, the samples could be divided into a western grouping and an eastern grouping on the basis of the red garnet and tremolite content.

Further retrievals were made as checks on the data in the file.

MATHEMATICAL ANALYSIS OF THE HEAVY MINERAL DATA

In order to assess the data in a more objective manner, numerical analytic techniques were used, which included tests to determine the frequency distribution of the data, R-mode factor analysis, and linear discriminant function analysis.

One of the main assumptions in parametric statistics is that the data are normally distributed. If they are not, then an attempt must be made to determine their frequency distribution. Appropriate transformation can often be made to normalize the data distributions.

The frequency distribution of each variable was compared to a normal distribution using the Chi-square test, using both the raw data and the logarithims (base 10 and base e) of the data. They were also tested as Poisson and positive and negative binomial distributions using the Chi-square test alone (Griffiths, 1960).

In transforming the data so that it approaches a normal distribution, it is important to avoid distorting the original data to the extent that a loss of information occurs. To this end, only linear transforms were applied. The transforms used were to take

TABLE 3

RETRIEVAL REQUEST USING BASIC IGNEOUS ROCKS AS THE CRITERION

IF   (('GBBR' = LITHOLOGY/I/) OR ('DIBS' = LITHOLOGY/I/)

          OR ('ANDS' = LITHOLOGY/I/) OR ('BSLT' = LITHOLOGY/I/) OR

          ('DORT' = LITHOLOGY/I/))

PUNCH   ID-NUMBER LOWER-GRAIN-SIZE/1/ WEIGHT-HEAVIES/1/

          PERCENT-MAGNETITE/1/ TOTAL-GRAINS/1/ HBLD/1/ BSHB/1/

          TRML/1/ RBCK/1/ OPRXN/1/ CPRXN/1/ AUGT/1/ RGRN/1/

          PGRN/1/ ZOIS/1/ EPDT/1/ RUTL/1/ SPHN/1/ ZRCN/1/ OPAQ/1/

          MICA/1/ TMLN/1/

IN FORMAT N/5/ N/1/ .N/3/ N/2/ .N/2/ N/2/ .N/2/ N/3/ N/3/ N/3/ N/3/

          N/3/ N/3/ N/3/ N/3/ N/3/ N/3/ N/3/ N/3/ N/3/*

          N/3/ N/3/ N/3/ STOP

Note: See Table 1 for list of abbreviations.

118

the natural logarithims of the data where appropriate ($\log_e(x+1)$ $\cdots$ and calculate Z-scores ($Z = \bar{x}/\sigma$). The first transformation converts a $\log_e$-normal distribution into a normal distribution, and the second standardizes the mean and variance of each variable to avoid mixing variances of transformed and untransformed variables.

R-Mode factor analysis is a mathematical procedure used to establish, if possible, relationships among a set of variables. It indicates whether the variance in an original set of variables can be accounted for adequately by a smaller number of multivariables termed factors. Mathematically, each factor is described by an eigenvector and the eigenvalue. Geometrically, the eigenvectors (factors) can be regarded as orthogonal axes (i.e. independent axes) which are fitted to the data in an N dimensional space, where N is the number of variables describing each sample. The resultant factors in this solution are principal component factors.

Once these factors have been computed (i.e. they replace the original variables) an iteritive procedure is used in which successively the axes are rotated, maintaining the orthogonality, so as to maximize the variance of the data set, i.e. so that the greatest percentage of the variance is explained by the factors. After each iteration, the factor which explains the least amount of variance is combined with its most closely associated factor to form a new factor. This rotation procedure uses the Varimax criterion.

As stated by Harman (1967) a principal objective of factor analysis is to obtain a parsimonious description of the observed data. The parsimony is accomplished, in a statistical sense, by the attempt to explain as much of the variance of each variable and of the total variance using as few factors as possible. Normally, it is desirable to explain at least 85 percent of the total variance, and 80 percent of the variance of each variable (termed communality) accounted for by the factors. Each factor then represents some combination of the original variables, and contains the same information as the original variables. However, the meaning of each factor is interpreted in the light of the variables composing the factors (factor loadings) and any other geologic information. The theory and descriptions of the technique and examples of its use are presented by Cameron, 1967; Garrett and Nichol, 1968; Harman, 1967; Hendrickson and White, 1864; and Klovan, 1968.

One use of R-Mode factor analysis is to determine groupings of samples on the basis of the strength of their association (factor scores) with a given factor. Using these *a priori* groups, a discriminant function analysis can be performed. This analysis determines the mutual exclusiveness of the groups, and calculates the probability of each sample belonging to each of the groups.

The discriminant analysis is based on linear functions (discriminant functions) which comprise a proportion of the original variables (mineral species). The more a variable assists in discriminating between groups, the greater weight it is given in the discriminant function. In this way, the analysis can also assist in deciding which minerals are most useful in distinguishing between glacial lobes. Finally, the technique can also be used to classify unknown samples. It calculates, on the basis of the discriminant functions, the probability of membership in a group.

NORMALITY TESTS

Eighteen variables (mineral data) were tested for normality using two computer programmes; SNORT (Preston, 1970) in which the main criteria is the Chi-square test, and HSTGM (R.A. Olson, Department of Geology, University of Western Ontario) which comprises several tests including Method of Moments.

The type of distribution for each variable applies to all three grain sizes studied (Table 4). The tests were applied first using a class interval of 0.25 standard deviations of each variable, and secondly, using a class interval of 0.125 standard deviations.
Minerals that were abundant in the samples had normal or $\log_e$-normal distributions. None of the minerals had binomial or Poisson distributions as determined by the Chi-square method (Ondrick and Griffiths, 1969).

R-MODE FACTOR ANALYSIS RESULTS

In view of the results of the retrievals from the data file, a geological model was suggested which indicates that heavy mineral assemblages are related to provenance. Grossly, there are two main sources of the assemblages: the Superior and Southern Provinces and the Grenville Province. Within the Superior and Southern Provinces three assemblages are present: a dominant assemblage reflecting the general low grade of a metamorphism in these provinces, another assemblage having a source in basic igneous rocks, and an assemblage with a source in fine grained sediments or mineralized areas. Within the Grenville Province two sub-sources can be distinguished: one in the Georgian Bay area between Sudbury and Lake Simcoe, the other east of Lake Simcoe.

TABLE 4
FREQUENCY DISTRIBUTION OF HEAVY MINERAL DATA AS TESTED BY MOMENTS METHOD[1]

MINERAL	NORMAL	LOG$_E$ NORMAL	UNDEFINED
WTHV[2]		***	
MGNT			***
HBLD	***		
BSHB		***	
TRML		***	
OPRX		***	
CPRX		***	
RGRN[3]		***	
PGRN[3]		***	
TGRN[4]		***	
ZOIS		***	
EPDT			***
RUTL			***
SPHN		***	
ZRCN			***
OPAQ		***	
MICA			***
TMLN			***

1- Class interval is 0.125 standard deviation of each variable.
2- See Table 1 for list of abbreviations.
3- In fine sand (0.25-0.125 mm) only.
4- In very fine sand and coarse silt classes (0.125-0.063 mm) only.

In the first run of the R-Mode analysis, 14 variables were used for all 115 samples from the Precambrian areas. The minerals excluded were basaltic hornblende, riebeckite, mica, and tourmaline because they are rare or absent in all but a few samples. A six-factor solution explains nearly 80 percent of the total variance although several of the communalities are below an acceptable level (0.8) (Table 5). In this first run, it is evident that certain variables (orthopyroxene, rutile, and zircon) are highly loaded in certain factors mainly because they are absent or present in only small amounts in the fine sand fraction. Also, it is thought that since hornblende and clinopyroxene are strongly negatively correlated (R= -0.65) and are grouped in the same factor, hornblende could be excluded along with the other three minerals. Briefly, the geological reasons for excluding hornblende are that it has more varied paragenises than clinopyroxene, and as a result, hornblende would be less indicative of a particular source than would clinopyroxene.

When the factor scores for each sample are examined, it is found that all but 10 of the samples can be assigned unequivocally to one of the factors: that is only 10 of the samples have factor scores which are different by less than 0.2 standard deviations. These samples were excluded in subsequent runs of the factor analysis. In the second run (based on ten variables and using 105 samples), the six-factor solution is acceptable (Table 5).

Clearly, there is considerable improvement in both the total variance explained and the communalities of the variables, all of which are above 80 percent. From a statistical point of view, the six factor solution is acceptable. They explain 88 percent of the variance. The four other potential factors account for only 12 percent of the variance, and as a result can be considered relatively less unimportant.

GEOLOGICAL INTERPRETATION OF FACTORS

Five of the factors can be interpreted as indicators of Provenance (Table 6). The high loading of red and purple garnets in Factor 1 is perhaps most typical of the area in the Grenville Province from north of Lake Simcoe northwest toward the Grenville Front (Fig. 1). Whereas the association of tremolite and purple garnets in Factor 6 is more typical of the Grenville area east from Lake Simcoe or thereabouts. Factors 2 and 4 are interpreted as reflecting sources in the Superior and Southern Provinces. Low weight of heavy minerals and high epidote (Factor 2) are expected in most areas of these Provinces, and opaque minerals (Factor 4) are thought to be derived principally from fine-grained sediments and

TABLE 5

RESULTS OF R-MODE FACTOR ANALYSIS (FINE SAND SIZE)

14 – VARIABLES

FACTORS	1	2	3	4	5	6
	+MGNT*	+WTHV	+TRML	+OPAQ	-OPRX*	+ZRCN
	-HBLD	+RGRN*	+ZOIS*		-PGRN	
	-CPRX	-EPDT	+RUTL			
	-RGRN*	+SPHN				
	-SPHN					

TOTAL VARIANCE = 78.2 percent.

10 – VARIABLES

FACTORS	1	2	3	4	5	6
	+CPRX	-WTNV	+ZOIS	+OPAQ	+TRML	+MGNT
	-RGRN	+EPDT	+SPHN		+PGRN	
	-PGRN					

TOTAL VARIANCE = 87.9 percent.

Notes:

*– indicates communalities less than 80 percent for each variable. Signs indicate positive or negative loading of variable in factor. See Table 1 for list of abbreviations.

TABLE 6

GEOLOGICAL INTERPRETATION OF FACTORS

FACTOR[1]	MINERAL[2]	
1	RGRN PGRN CPRX	Grenville Province: western source (GW)[3] (Georgian Bay area ?).
2	WTHV EPDT	Superior-Southern Provinces in general (SS).
3	SPHN ZOIS	Unknown, perhaps multiple source (U).
4	OPAQ	Generally fine-grained sediments, e.g. iron formation, mineralization. Mainly Superior-Southern Provinces (SM).
5	TRML PGRN	Grenville Province: eastern source (GE).
6	MGNT	Source of magnetic minerals, e.g. basic igneous rocks, iron formation, certain metamorphic rocks (M).

1- Based on 10 variables
2- See Table 1 for list of abbreviations.
3- These abbreviations are used in Tables 7 and 9.

124

perhaps mineralized areas. A strong loading of the magnetic fraction in Factor 6 is interpreted as indicating basic rocks and certain metamorphic rocks high in magnetite.

The association of zoisite (and clinozoisite) and sphene in Factor 3 is enigmatic. One would have expected epidote, zoisite and clinzoisite to be associated in one factor, and sphene associated possibly with other Grenville source indicators, as alkaline intrusions in which sphene is common are more abundant in the Grenville Province (e.g. Bancroft area). Alternatively, the assocation of a sample with Factor 3 may be an indication that the sample has multiple sources. Since these minerals are not strong indicators of a particular source, they might be highly loaded in this factor by default so to speak. In this regard, it is observed that several samples that could not be definitely assigned to one factor had similar factor scores in Factor 3 and one of the other Factors. This, perhaps, is a further indication of a multiple source of these samples.

The result of the factor analysis for the two finer grain sizes (very fine sand and coarse silt) are statistically the same as those for fine sand. However, a geological interpretation of the factors is made impossible by the association of minerals in two of the factors which are geologically meaningless. Further, on examining the factor scores, many of the samples have factor scores of equal value. It is concluded that because no distinction can be made between the garnet colours in the finer sizes, numerical techniques cannot be successfully applied to the data in these sizes.

LINEAR DISCRIMINANT ANALYSIS RESULTS

On the basis of the factor scores there is a definite grouping of samples into five sub-groups, which can be related to different provenances. A discriminant analysis was made to determine the probability of group membership and as a further check on the existence of the five groups. The samples were divided into five groups (a priori groups) on the basis of their factor scores, and these were used to calculate linear discriminant functions. All the samples from the Precambrian area which could not be clearly assigned to one of the factors were excluded from the a priori groups and classified later using the discriminant analysis. Ninety-four samples made up five a priori groups.

In the first run of the analysis, only three of the ninety-four samples had probabilities of group membership less than 90. Of the remaining samples, 78 percent had probabilities greater than 95.

The three doubtful samples were reclassified as unknowns. Hoping to increase the discrimination between the groups and so increase the ability to classify the unknown samples, the discriminant analysis was rerun with 91 samples among the five groups. Results from this run are that 85 percent of the samples have greater than 95 percent probability of group membership as assigned. From this it is concluded that there are indeed five groups that can be related to five provenances.

The classification of the doubtful (undecided) samples from the Precambrian area are presented in Table 7. It is seen that a sample assigned to one of the two main source areas was consistently correctly classified by the numerical methods. About half of the samples which were assigned to one of the five sub-sources on the basis of visual examination were correctly assigned by the numerical techniques. The remaining samples show the effect of the multiple sources (e.g. samples 4108, 0911, 4104, 4102, 4163). Still other samples indicate that local sources probably contributed to the till, such that assignment to a source area other than in a broad category is impossible. For example, samples 4142, 4148, and 4149 are all in the Grenville area, however, they have characteristics which perhaps indicate some kind of transitional zone in the bedrock, or alternatively, markedly different lithologies in the area.

The discriminant analysis showed that the following minerals or mineral groups were most useful in distinguishing between groups: weight of heavy minerals, magentic fraction, tremolite, clinopyroxene, purple and red garnets, epidote, and opaque minerals.

CLASSIFICATION OF SAMPLES FROM PALEOZOIC TERRAIN

Seventeen samples were analysed and used to test the results of this study. The samples come from the Paleozoic terrane in southern Ontario, and eastern and northern Michigan. The samples were classified in terms of the glacial lobes that transported and deposited the tills, location and other field evidence, graphical examination of the heavy mineral data, discriminant analysis, and the content of the elements Cu, Ni, Zn, Cr, CaO (May, 1971) (Table 9).

Four of the source areas (factors) can be associated with three glacial lobes (Table 8). It is suggested that the western Grenville area is the source of the main characteristics of the Georgian Bay lobe. However, more work would have to be done to substantiate this. For the present, the western Grenville source is referred to as

TABLE 7

COMPARISON OF CLASSIFICATIONS OF SAMPLES FROM PRECAMBRIAN AREAS (FINE SAND)[1]

SAMPLE NUMBER	STRUCTURAL PROVINCE	GRAPHICAL METHOD	FACTOR ANALYSIS	DISCRIMINANT ANALYSIS (% Probability)
2526	SU	SS	SS, U	SS (98)
4126	SU	SS	SS, GE	SS (89), GW (7)
4125	SU	SS	SS, GE	SM (97), SS (3)
4108	SO	M, SS	SS, M	M (54), SS (46)
0911	SO	SS	SS, SM	SS (91), SM (9)
4104	GR	M	SM	SM (77), GW (15), GE (6), SS (2)
4148	GR	GW	GW, GE	GW (86), SS (14)

127

TABLE 7 (cont.)

SAMPLE NUMBER	STRUCTURAL PROVINCE	GRAPHICAL METHOD	FACTOR ANALYSIS	DISCRIMINANT ANALYSIS (% Probability)
4149	GR	GE	GE	GW (69), SS (26), GE (4)
1881	GR	M, GE	M, GE	GE (85), SS (13)

1- These samples were associated weakly with one or more factors in the factor analysis.

2- Structural Provinces.

    SO - Southern Province   SU - Superior Province
    GR - Grenville Province

3- SS - Superior and Southern Provinces in general.
    M - Source of magnetic minerals, basic igneous and metamorphic rocks.
    GW - Western area in Grenville Province.
    GE - Eastern area in Grenville Province.
    SM - Source of opague minerals, fine-grained sediments or mineralization.
    U - Unknown or multiple sources.

128

TABLE 8

RELATIONSHIP OF LOBES TO SOURCE AREAS

Huron lobe   Superior-Southern Province indicators.

                    Magnetic mineral source.

                    Fine grained sediments and mineralization.

Ontario lobe   Eastern Grenville Province source with possible influence

Erie lobe   from western Grenville area.

Georgian Bay lobe   Western Grenville Province source.

TABLE 9

CLASSIFICATION OF SAMPLES FROM PALEOZOIC AREAS

Southern Ontario

SAMPLE[1] NUMBER	FIELD CRITERIA	HEAVY MINERALS	DISCRIMINANT ANALYSIS (% Probability)	GEOCHEMISTRY[2]
1	H[3]	H	H (56), G (43)	–
2	H	H (G?)	H (99)	–
3	O	O	O (97)	–
4	O	O	O (75), G (25)	–
5	H	H	H (99)	–
6	O	O	O (99)	–
7	H	H	H (100)	H and G
8	H	H	H (99)	H
9	O	O	O (100)	E-O
10	G	O-G	O (88), G (12)	G (?)

TABLE 9 (cont.)

SAMPLE NUMBER	FIELD CRITERIA	HEAVY MINERALS	DISCRIMINANT ANALYSIS (% Probability)	GEOCHEMISTRY	
				Southern Ontario	Eastern Michigan
11	O	O	O (99)	O	
12	H or E	H and E	H (74), E (25)	E (mixed H and E)	
13	H or G	G, O-E	O-E (99)	H or G (?)	
14	H	H (G)	H (99)	—	
15	H	H (G)	H (99)	—	
16	H	H (G)	H (99)	—	
17	H	H (G)	H (99)	—	

1- See Fig. 1 for locations.

2- After May, 1971.

3- H – Huron lobe   G – Georgian Bay area (lobe?)

    O – Ontario lobe   E – Erie lobe.

the Georgian Bay area.

As seen in Table 9, all but four samples can be clearly assigned to a single lobe. Samples 4182 and 4184 reflect two sources, and they are both in an interlobate area between an eastern source and a northern source. Sample 1029 reflects mixing of the Huron lobe with ice crossing the Georgian Bay area as proposed by Dreimanis et al. (1957). Sample 2330 was most likely deposited by the Ontario lobe but it appears to have in addition incorporated material transported southward by the Georgian Bay lobe.

In this study it is not possible to distinguish between lobes in Lake Superior and Lake Michigan basin. This would require further sampling in northern Michigan and Wisconsin and in Minnesota. It is possible, however, that till deposited by a Superior lobe would reflect both the basic bedrock and fine grained sediments and mineralization that are abundant in that area (Fig. 1), whereas tills deposited by the Michigan lobes might have a general Superior-Southern assemblage.

SUMMARY AND CONCLUSIONS

In this case history, a relatively small computerized information file has been generated using SAFRAS. The file comprises the identification and location of 132 till samples, the age and lithology of bedrock in their vicinity, and point count data on 18 heavy minerals and mineral groups for each sample. Reduction of this information into computer processible form was found to be efficient and relatively error free.

Retrievals from the file using several criteria suggest a geological model in which five mineral assemblages appear to be related to provenances in two main areas: the Superior and Southern Provinces and the Grenville Province of the Canadian Precambrian Shield.

Two mathematical techniques have been used to determine the numerical basis of the empirical model: R-Mode factor analysis and discriminant analysis. The result is that the samples can be separated into five groups on the basis of their mineral assemblages, and these can be interpreted geologically as reflecting particular areas of provenance. In turn the areas of provenance are related to specific glacial lobes that deposited the tills in southern Ontario and eastern Michigan. Testing of this relationship using samples from these areas shows that the Huron, Erie, Ontario and probably the Georgian Bay lobes can be distinguished on the basis of the heavy mineral content of the tills they deposited.

REFERENCES

Cameron, E.M., 1967, A computer program for factor analysis of geochemical data: Geol. Surv. Can. Paper 67-34, 42 p.

Garrett, R.G., and Nichol, I., 1968, Factor analysis as aid in the interpretation of regional geochemical stream sediment data: Quart. Colorado School Mines, v.   , pp. 245-264.

Griffiths, J.C., 1960, Frequency distributions in accessory mineral analyses: Jour. Geol., v. 68, pp. 353-365.

Harman, H.H., 1967, Modern factor analysis. Chicago Univ. Press, 2nd ed., 474 p.

Hendrickson, A.E., and White, P.O., 1964, Promax: a quick method for rotation to oblique simple structure: British Jour. Stat. Psych., v. 17, pp. 65-70.

Klovan, J.E., 1968, Selection of target areas of factor analysis: Proc. Symposium on Decision-Making in Mineral Exploration, Jan. 1968, Vancouver, B.C., Western Miner Press, pp. 19-27.

May, R.W., 1971, Cu, Cr, Ni, Zn, CaO, MgO content of Wisconsin tills in southern Ontario: Ph.D. thesis, Department of Geology, University of Western Ontario, 189 p.

Preston, D.A., 1970, Fortran IV program for sample normality tests: Computer Contribution 41, State Geological Survey, The University of Kansas, Lawrence, Kansas, 27 p.

Vagners, U.J., 1969, Mineral distribution in tills in central and southern Ontario: Ph.D. thesis, Department of Geology, University of Western Ontario, 277 p.

THE APPLICATION OF LINEAR DISCRIMINANT ANALYSIS TO THE INVESTIGATION OF TILLS

Ronald W. May

Introduction

Classification procedures are of two general types:
1) those utilized when there is no previous (*a priori*) knowledge of the natural groupings of a set of data.
2) those utilized when natural groupings are known to exist.

Type 1 procedures, referred to generally as "cluster analysis" can be employed to analyze both quantitative (e.g. Klovan, 1966) and semi-quantitative data (e.g. Bonham-Carter, 1965), whereas type 2 procedures usually require quantitative data. The latter technique permits discrimination between two or more groups of data on the basis of a linear combination of the variables measured and is thus known as "linear discriminant analysis".

Cluster Analysis

It is perhaps desirable to discuss this aspect of classification techniques very briefly for reasons which will become evident.

Basically "cluster analysis" is the name given to a variety of procedures which attempt to place together in a group or "cluster" those samples which are more similar to one another than to samples in other groups. A typical situation is shown in Fig. 1. In this illustration there is no *a priori* knowledge of the natural categories which exist. A cluster analysis might produce the results shown in fig. 2. - the existence of two groups with a slight overlap between them. A problem arises when another sample is taken and it is desired to place it into one of the groups. The cluster analysis must be performed again. However, there is a solution to this problem as outlined below.

Linear Discriminant Analysis

If the characteristics of the natural groups are known or can be resolved by the use of clustering methods, then a discriminant function can be derived which will permit the classification of any unknown samples into one of the various groups. The attributes which are used must be quantitative, i.e. measurable on an interval or ratio scale (e.g. feet, metres, per cent).

Discriminant analysis helps to determine the optimum way to distinguish between *a priori* established groups on the basis of the

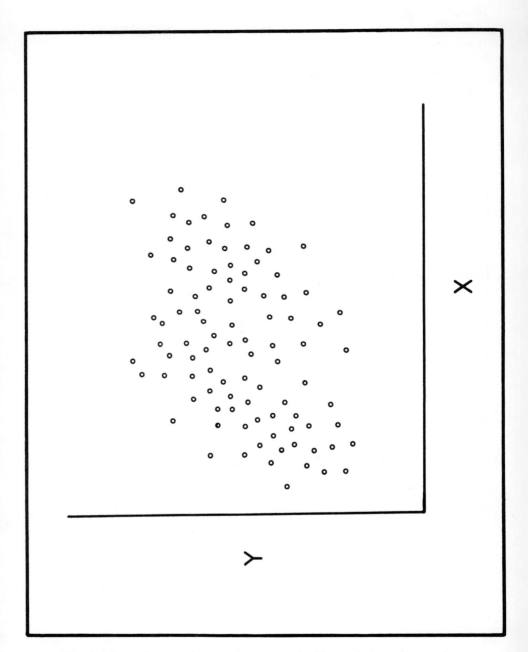

Fig. 1   Raw data unclustered — no *a priori* knowledge of groupings.

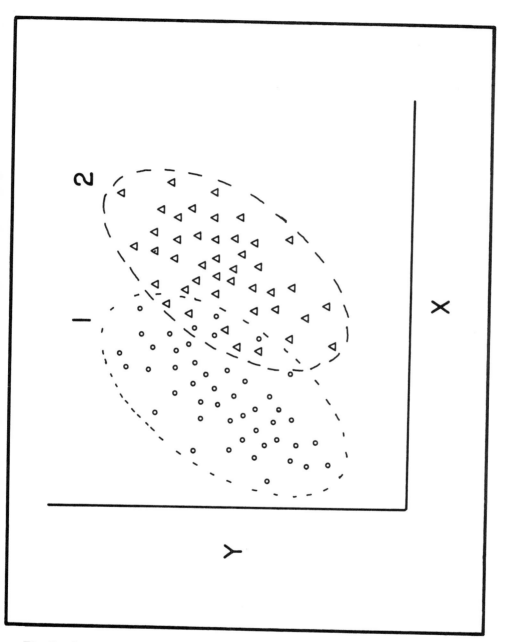

Fig. 2 Raw data after cluster analysis — two overlapping groups present.

measured variables. It also provides criteria for deciding which of these variables are the best for differentiating the various groups. For example, fig. 3 shows two groups with identical Y mean values and different X mean values. The arrangement of the groups in the two-dimensional space is such that the minimum overlap between the groups occurs along the X-axis. The "optimum way" to discriminate between groups 1 and 2 in this case is by using their X value. Thus, if a discriminant function were to be calculated for this example then it would have the form

$$D_i = aX_i + 0.0Y_i \qquad \qquad ...(1)$$

In this example the Y-value makes no contribution to the discriminant function.

Fig. 4 shows two groups with identical Y mean values and different X mean values. In this case the arrangement of the group ellipsoids is such that the "optimum way" to discriminate between the two groups is to use a combination of X and Y, specifically an equation of the form

$$D_i = aX_i + bY_i \qquad \qquad ...(2)$$

This equation is formulated by constructing two lines perpendicular to each other (denoted lines I and II in fig. 5) such that the region of overlap between the two groups along the line II is a minimum. Line II is the discriminant function. When an X and Y value are substituted into the equation ((2) above) a discriminant score D_i is obtained for each sample.

Classification of unknown samples into group 1 or 2 involves the use of a decision rule. The X and Y values of the point where line I intersects line II are substituted into equation (2) and the value obtained (denoted D_o) is used in the formulation of the decision rule. Basically if the D_i value for the unknown is less than D_o then it is classified in Group I, and conversely, if D_i is greater than D_o, then the unknown is placed in group 2. Since there is a "region of overlap" between the two groups, some of the unknown samples may be misclassified. It is thus necessary to obtain an estimate of the probability of misclassification. This is done by calculating the discriminant scores for the known samples and determining the percentage of these which are misclassified by the decision rule. This "percentage" gives the "probability of misclassification" for an unknown sample.

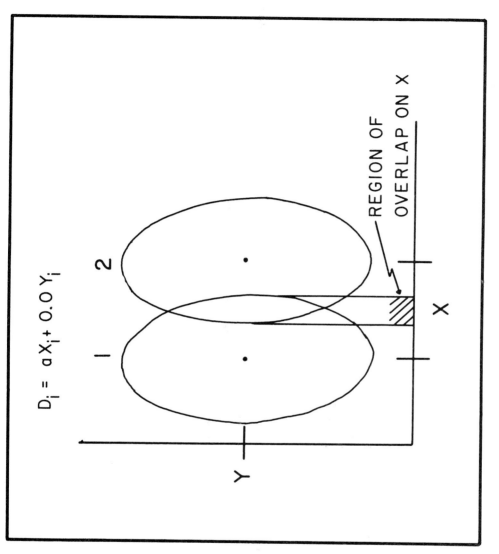

Fig. 3 Example of discriminant analysis, two groups have same Y-mean values, X is the only variable which contributes to the differentiation of the two groups.

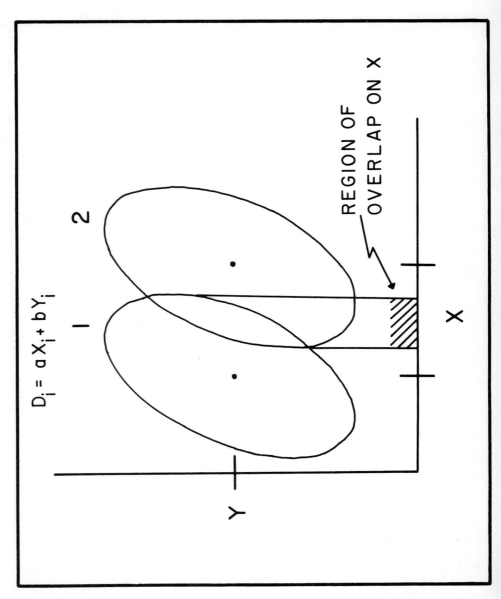

Fig. 4 Discriminant analysis, two groups have similar Y-mean values, but arrangement groups is such the overlap on X is greater than in Fig. 3.

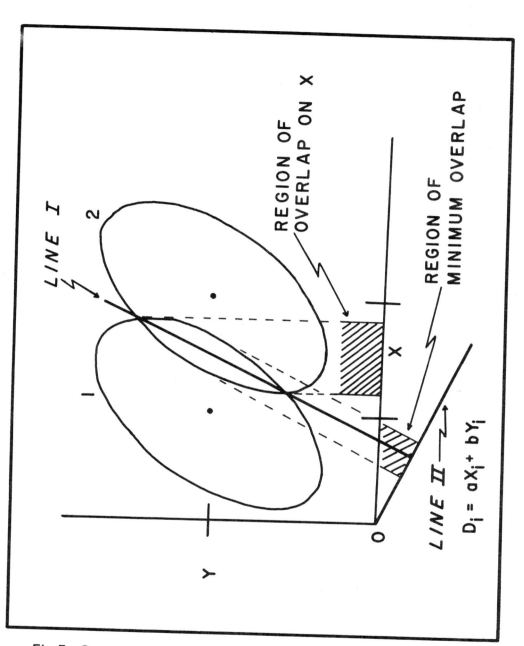

Fig. 5 Geometric construction of discriminant function $D = aX_i + bY_i$.

Application of Discriminant Analysis

Discriminant analysis was developed originally from research dealing with the use of multiple measurement data in taxonomic and biometric problems. Barnard (1935), Fisher (1936, 1938). Discussion of the utilization of this technique in the geosciences is given by Griffiths (1966) and Klovan and Billings (1967). The illustration provided here, of its use in the investigation of tills is based on a study of till geochemistry in southern Ontario (May, 1971).

Table 1 gives the means and standard deviations for Cu, Zn, Cr, Ni, CaO and MgO, for the three groups of samples studied. Although the Ontario lobe averages are more similar to those of Erie lobe than to the Huron-Georgian Bay lobe (henceforth referred to as the Northern group) the large standard deviations of the Ontario lobe groups makes its inclusion with the Erie lobe somewhat premature. To investigate the reasons for the large variability of the Ontario lobe samples, it was decided to use discriminant analysis techniques.

The Erie and Northern groups were used as the known or *a priori* groups and the Ontario lobe samples were treated as "unknowns". Utilizing all variables as a first approximation a discriminant function

$$D_i = 0.34\,Cu - 0.07\,Zn - 0.02\,Cr + 0.40\,Ni - 0.03\,CaO - 0.83\,MgO \quad \text{...(3)}$$

was obtained. CaO and MgO were eliminated to test the feasibility of trace elements only to differentiate the two groups. Zinc was also eliminated because the distribution was not normal (an assumption inherent in most multi-variate statistical techniques). Discriminant analysis yielded the function

$$D_i = 0.58\,Cu - 0.02\,Cr + 0.84\,Ni \quad \text{...(4)}$$

Histograms of the discriminant scores for the known groups and those of the Ontario lobe group are shown in Fig. 6. The D_o value which best separates the two known groups is 37. A possible reason can now be suggested for the high variability of the Ontario lobe samples. They appear to have affinities with both of the known groups. An investigation of the areal distribution of the Ontario lobe samples reveals that those with a

TABLE 1

Cu, Zn, Cr, Ni, CaO, MgO MEANS AND STANDARD DEVIATIONS FOR ERIE, HURON–GEORGIAN BAY AND ONTARIO LOBE SAMPLES

		Cu	Zn	Cr	Ni	CaO	MgO
Erie Lobe	$\bar{x}$	25	64	61	38	16.6	3.9
N=27	s	3	7	16	5	1.1	0.6
Huron–Georgian Bay Lobe	$\bar{x}$	20	55	50	24	18.1	5.8
N=42	s	4	10	14	5	4.4	1.6
Ontario Lobe	$\bar{x}$	29	67	49	31	16.1	4.2
N=34	s	10	27	16	14	6.5	3.0

Note: Cu, Zn, Cr, Ni values are in ppm.
CaO, MgO values are in per cent.
N = no. of samples

143

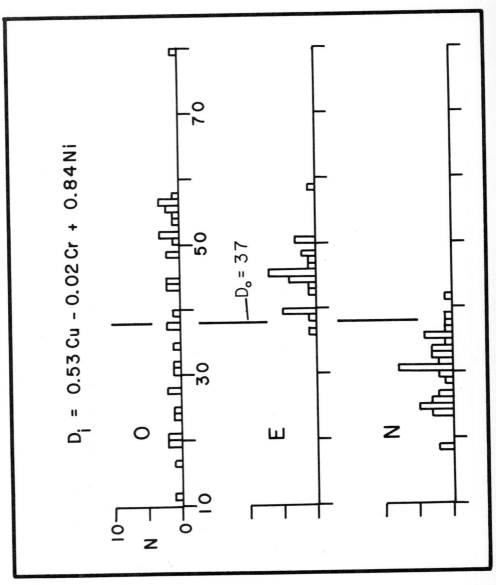

Fig. 6 Histogram of discriminant scores of Erie lobe (E), Huron-Georgian Bay lobe (N) and Ontario lobe (O) samples.

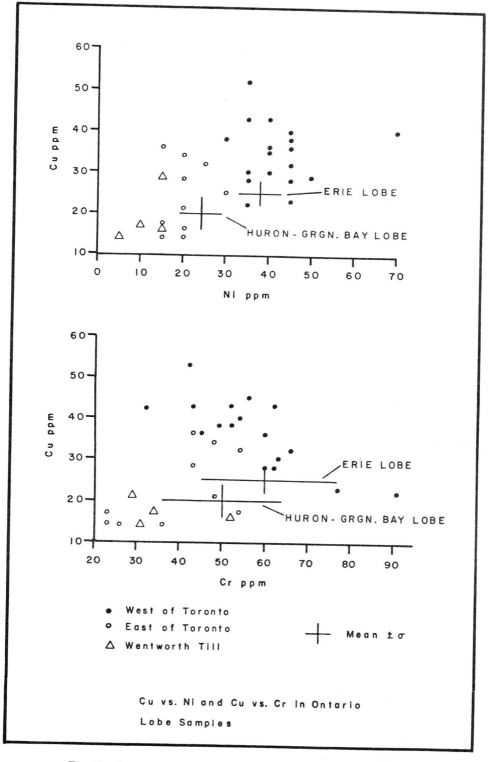

Fig. 7 Cu vs. Ni and Cu vs. Cr in Ontario Lobe Samples.

Northern affinity are of two types: (1) Wentworth Till at the northwestern extremity of the Ontario lobe, and (2) those samples which were taken east of Toronto. Samples with Erie lobe characteristics were taken east of Toronto and the Wentworth Till area. The existence of these subgroups within the Ontario lobe group is illustrated in Fig. 7 showing Cu against Ni and Cr. An examination of the bedrock distribution reveals that these samples west of Toronto are on shale bedrock whereas those east of Toronto are not. The only exceptions west of Toronto are four samples of Wentworth Till which are unlike other samples in the same area in that they are classified by the discriminant analysis, with the Huron-Georgian Bay lobe group. Karrow (1963) has shown that the Wentworth Till was deposited by a glacier moving northwest out of the Lake Ontario basin. The conflict between the published interpretation and the assignment on the basis of geochemical criteria can be resolved if it is postulated that the glacier incorporated material derived from a northern source (i.e. Huron-Georgian Bay lobe). That the till does overlay outwash has been stated by Karrow (1963); however, it is not known to which till sheet the material is related. Vagners (1966) has suggested on the basis of pebble lithologics, that in the Kitchener area the Wentworth till overlies outwash from a northern or northeastern source. The interpretation based on geochemical data is consistent with the latter explanation.

CONCLUSION

The example used here, although utilizing till geochemistry, could have been based on any other measurable quantities. For example, any or all of pebble lithology, carbonate content of the till matrix, granulometric composition, mineralogy and engineering properties could be used in studies of a similar nature. The application of linear discriminant analysis to till geochemistry has enabled the demonstration of a relationship between shale bedrock and the trace element content of glacial tills in southern Ontario. The incorporation of material derived from a different source has also been shown.

A rigorous approach to discrimination and classification can be of value in relating measurable variables to geological processes.

ACKNOWLEDGEMENTS

Drs. A. Dreimanis and P. Sutterlin are thanked for their discussion and comments on the manuscript. Financial support for this study came from National Research Council of Canada grant A4215 to Dr. Dreimanis.

REFERENCES

Barnard, M.M., 1935, The secular variations of skull characters in four series of Egyptian skulls, Ann. Eugen., Vol. 6, pp. 352-371.

Bonham-Carter, G.F., 1965, A numerical method of classification using qualitative and semi-quantitative data, as applied to the facies analysis of limestones, Bull. Can. Petrol. Geol., Vol. 13, No. 4, pp. 482-502.

Fisher, R.A., 1936, The use of multiple measurements in taxonomic problems, Ann. Eugen., Vol. 7, pp. 179-188.

Fisher, R.A., 1938, The statistical utilization of multiple measurements, Ann. Eugen., Vol. 8, pp. 376-386.

Griffiths, J.C., 1966, Application of discriminant functions as a classification tool in the geosciences, in Merriam, D.F., ed., Computer Applications in the Earth Sciences: Colloquium on Classification Procedures, Contribution 7, State Geologocial Surv., The University of Kansas, pp. 48-52.

Karrow, P.F., 1963, Pleistocene geology of the Hamilton-Galt area, Ont. Dept. of Mines, Geol. Report 16, 68 pages.

Klovan, J.E., 1966, The use of factor analysis in determining depositional environments from grain size distributions, Jour. Sed. Petrol., Vol. 36, No. 1, pp. 115-125.

Klovan, J.E. and Billings, G.K., 1967, Classification of geological samples by discriminant function analysis, Vol. 15, No. 3, pp. 313-330.

May, R.W., 1971, Cu, Ni, Zn, Cr, Cao and MgO content of Wisconsin tills in southern Ontario, unpublished Ph.D. dissertation, University of Western Ontario, London, 188 pages.

Parks, J.M., 1966, Cluster analysis applied to multivariate geologic problems, Jour. Geol., Vol. 74, No. 5, Pt. 2, pp. 703-715.

Vagners, U.J., 1966, Lithologic relationship of till to carbonate bedrock in southern Ontario, unpublished M.Sc. thesis, University of Western Ontario, London, 154 pages.

USE OF Q-MODE FACTOR ANALYSIS IN THE INTERPRETATION OF GLACIAL DEPOSITS

Allan Falconer

Glacial deposits in most areas of the world have been modified as a result of post glacial climate changes. Whilst this is generally recognised, many studies of glacial deposits exist without measures of these modifications or any safeguards against misinterpretation of them. In the study reported here an attempt is made to introduce some objective measures into the interpretation of the suite of deposits existing in a glaciated upland valley of north-eastern England.

The study area, as outlined in Figure 1, is the upper part of the valley of the River Wear referred to below as Upper Weardale. All previous studies of this area support the theory that the valley was glaciated but there is some dispute about the extent of the ice cover and the significance of the distribution of the deposit types. As in most studies of such areas, the extent of the ice cover has been inferred from the distribution of various deposits or from the rock fragments in deposits.

A summary of the conclusions reached by previous workers is presented in Figure 2. All investigators agree that there are no erratics in Weardale which originate beyond the outcrop of the carboniferous rocks of the Northern Pennines. The map compiled by Eastwood (Figure 3) is a summary of the evidence provided by erractics up to 1953. In this map Upper Weardale is within the area defined as the Alston Block and this is bounded by the symbol indicating that the area was not invaded by ice from other centres. Dwerryhouse (1902) in his examination of the Weardale landscape concludes that the watershed areas in Upper Weardale remained ice-free and were thus nunataks at the period of maximum glaciation. This opinion is endorsed by the more recent work by Maling (1955).

The basis for this conclusion, in both cases is the distribution of glacial till in Upper Weardale, this is illustrated in Figure 4 which is based on the mapping done by Maling. Both Maling and Dwerryhouse comment on the nature of the till terming it a dense blue-grey clay with stones. The stones are typically striated and of the distinctive pentagonal form of till stones. Maling notes some similarities between this deposit and deposits elsewhere on the slopes of the region.

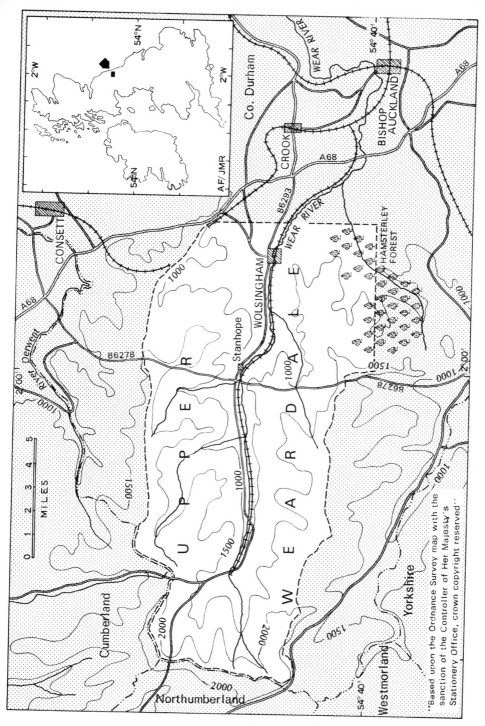

Fig. 1 The location of Upper Weardale.

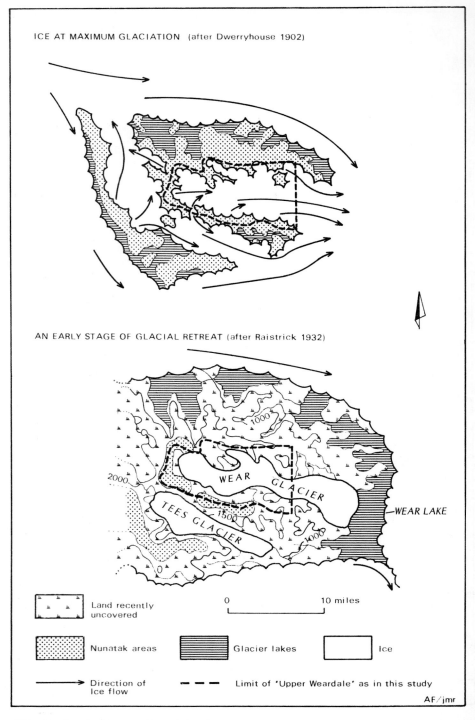

ICE AT MAXIMUM GLACIATION (after Dwerryhouse 1902)

AN EARLY STAGE OF GLACIAL RETREAT (after Raistrick 1932)

1000

2000

WEAR GLACIER

TEES GLACIER

1500

1000

0

WEAR LAKE

Land recently uncovered

0   10 miles

Nunatak areas

Glacier lakes

Ice

Direction of Ice flow

Limit of 'Upper Weardale' as in this study

AF/jmr

Fig. 2   Ice in Upper Weardale. Conclusions of Dwerryhouse and Raistrick.

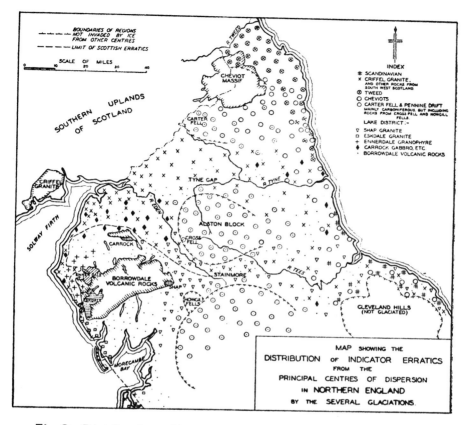

Fig. 3   Distribution of Indicator Erratics after Eastwood (1953).

151

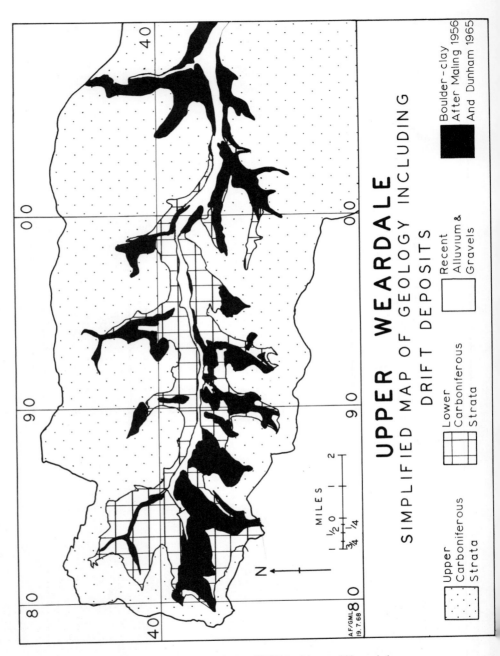

Fig. 4   Distribution of Till in Upper Weardale.

Evidence of glacial action when assessed in this manner, can, over large areas be of considerable value. However, investigations in Upper Weardale revealed that mapping the occurrence of till was not a simple task. In certain areas surficial deposits are not recorded by the Geology Survey map-makers because they are less than 3 feet in thickness. Maling objects to this but does not produce maps of the material which is less than 3 feet in depth. Thus the map presented as Figure 4 is a simplified record of the geology of the study region.

Upper Weardale is a valley which opens to the east but the valleys adjacent to it, open to the north and the south, namely the valleys of the East Allen and the Upper Tees. All of these river valleys are cut into the same stratigraphic sequence as Upper Weardale. This sequence is one of interbedded sandstones, shales and limestones typical of the cyclic deposition found in the carboniferous rock of northern England, the particular series here being the Yoredale series.

The nature of the rocks of the Yoredale series requires little comment. Figure 5 presents the stratigraphic column for Upper Weardale. The Lower Carboniferous (Yoredale Series) includes limestones which are dark-blue in colour, finely grained and thinly bedded. They are formed of calcite mudstone in which organic matter occurs as a dark pigment (Johnson 1963). The shales occuring with these limestones are usually dark-grey, hard, well-bedded and highly fossiliferous. Overlying these are ferruginous shales grading upwards into sandstones. Sandstones in Upper Weardale in both the Upper and Lower Carboniferous strata are either white or brown rocks, with sub-angular quartz grains 0.3-0.1 mm in diameter.

The succession of strata shown in Figure 5 is present over the area of the Alston Block and dips eastwards at about 130 feet per mile. The whole area has been faulted but the only significant displacement is along the line of the Burtreeford Fault. Associated with the faulting there are a large number of intrusions, dykes and sills with their associated metamorphism which introduces mineral veins and quartz dolerite into the region.

Because the adjacent areas are of similar geologic structure it is doubtful if rock fragments transported into Weardale from these areas could be detected. Therefore the reasoning of Dwerryhouse (1902) and Maling (1955) is not necessarily correct. It is equally doubtful that an ice sheet reaching more than 200 miles south of this region at its maximum extent would not over-ride the

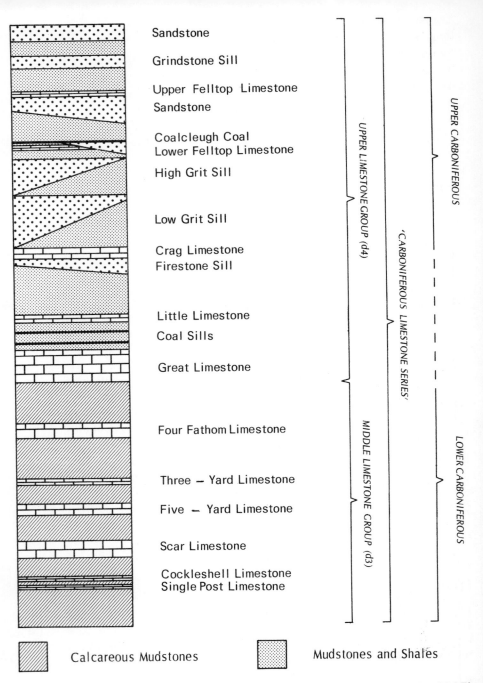

Sandstone

Grindstone Sill

Upper Felltop Limestone
Sandstone

Coalcleugh Coal
Lower Felltop Limestone

High Grit Sill

Low Grit Sill

Crag Limestone
Firestone Sill

Little Limestone
Coal Sills

Great Limestone

Four Fathom Limestone

Three — Yard Limestone

Five — Yard Limestone

Scar Limestone

Cockleshell Limestone
Single Post Limestone

UPPER LIMESTONE GROUP (d4)

'CARBONIFEROUS LIMESTONE SERIES'

UPPER CARBONIFEROUS

LOWER CARBONIFEROUS

MIDDLE LIMESTONE GROUP (d3)

Calcareous Mudstones

Mudstones and Shales

Fig. 5 Stratigraphic column for Upper Weardale (based on Dunham 1965).

comparatively low (2000 feet) watershed areas of Upper Weardale.

A recent study of the soils of Upper Weardale (Atkinson 1968) indicated that three major groups of parent material existed. In addition to these he considers the anthropogenic "spoil heaps" created by lead mining operations in the seventeenth to twentieth centuries but these are excluded from the present study. The three major types of parent materials were considered to be, upland regolith on ridges and interfluvial crests; solifluction deposits on slope flanks and valley sides; till and riverine alluvia in valley bottoms (Atkinson 1968 p. 29).

In further consideration of these parent materials Atkinson also comments; "Whilst the possibilities for polymorphism are substantial and all (parent materials) in fact may be the present day expressions of a single genetic feature (e.g. a Saale Till Sheet), each has received distinctive fashioning in the (post glacial) geomorphic history". (Atkinson 1968 p. 29).

Thus the present study is undertaken to evaluate the characteristics of the suite of deposits existing in a region where provenance is not a criterion for the differentiation of deposit types. The deposits which exist in Upper Weardale are generally recognised to have a complex genesis and yet an essentially similar character (attributable to the similarities in the source areas). As a notable comment on these deposits Atkinson concludes that, "It is extremely difficult to map the boundaries of regolith, solifluction deposits and till ..." (Atkinson 1968 p. 59).

The situation, therefore, is one in which different deposit types are identified although the mapping of these deposits is complicated by the transitions which exist between each of the major types. Thus the solifluction parent material described by Atkinson is stated to have "... many of the attributes of a glacial till which has undergone considerable congeliturbation". (Atkinson 1968 p. 47). In an attempt to evaluate the contribution of different processes to the genesis of the deposits in Upper Weardale it is necessary to clarify the nature of the conclusions reached by previous workers.

The presence of ice in Upper Weardale and the extent of it is inferred from maps of till distribution and the existence of striae and striated till stones (Dwerryhouse 1902; Maling 1955). The deposit-types described by previous investigators are notable for textural variability and sedimentary structures. Maling for example notes that ... "clay may exist ... right to the summit of the moors. Examination of this clay has shown that it is often sandy in texture". (Maling 1955 p. 88).

He also notes ...''other clays may approximate to boulder clay (till) in texture, but they are either devoid of erratics or, if stones are present, these are often sub-angular and do not show recognisable striae''. (Maling 1955 p. 88). and

''It can be argued, however, that the sandy clay irrespective of its origin, has been deposited upon the true glacial drift by later solifluction, soil creep or land-slip''. (Maling, 1955, p. 88).

Implicit in this work is the concept of a succession of processes operating through time. The characteristic deposit produced by each process, seemingly can be defined but the degree of alteration of one deposit type by some subsequent process can only be subjectively evaluated. Thus only the characteristic deposit types may be meaningfully evaluated unless there is some method of assessing the influence of each process on each sample. A model of this hypothesis may conveniently be defined as

$$\Sigma_\Phi = A\Sigma_\Phi{}' + B\Sigma_\Phi{}'' + C\Sigma_\Phi{}''' \qquad\qquad ..(1)$$

where Σ_Φ is the total sediment at any point and A. B. C, ... are coefficients identifying the contributions of Φ', Φ'' etc. to the sediment where Φ', Φ'' ... are assumed to represent the processes acting. The problem remaining is to reduce the sediment at any one point to some mathematical expression of properties.

In geomorphology the characteristic sediments produced by the action of a single process are frequently characterised by their texture. This type of analysis is further refined and presented by sedimentologists as a particle-size curve. That particle-size data may be used to discriminate between differing environments of deposition has been demonstrated at various levels by many sedimentologists (see Klovan 1966, Solohub and Klovan 1971, Krumbein and Pettijohn 1938, Krumbein and Graybill 1965 etc..). In the absence of differences in lithology, provenance of heavy minerals etc. it appears that the use of particle size data in this case may offer a suitable means of assessing the relative contributions of differing processes to the genesis of each deposit.

Following the method adopted by Klovan (1966) the sediment samples were analysed by seiving and hydrometer techniques to determine their particle size distribution. This was then prepared as a series of size classes each of 1.5 phi units, so breaking the data into 10 size classes. The data were gathered in two ways, in one case a purposive sample (defined after Krumbein and Graybill 1965) was taken and analysed, in a second case a random sample was taken and analysed. The results generated were clearly consistent and the results presented here are the results generated from the analysis of the random and purposive samples taken together as a third analysis.

The initial data formed a 150 x 10 matrix representing 150 particle-size distributions presented in 10 size classes, each class may be considered an independent variable. In such a case (Klovan 1965), Q-mode factor analysis of the data should reveal similarities between particle-size distributions which are derived from the individual sediment samples. Thus the factor analysis model

$$C = a_1 F_1 + a_2 F_2 + a_3 F_3 + a_4 F_4 \dots + e \qquad \qquad \dots (2)$$

has the same structure as the model implicit in the conclusions of previous workers. If particle size data for each sample (C) can be restated in terms of factors (F_1, F_2, F_3 ...) each of which has a coefficient (a_1, a_2, a_3 ...) for each sample plus some error term (e) then these factors, mathematically derived from the particle-size data should correspond with the effect of depositional processes (or groups of depositional processes) if these processes really do impart distinctive characteristics to a sediment as Klovan's (1966) results suggest. Furthermore the coefficients a_1, a_2 etc. should indicate the relative importance of the factors in the explanation of the sediment sample.

Table 1 lists the original data matrix and Table 2 lists the five factor solution. This solution is derived using a Q-mode factor analysis program based on a 'varimax' rotation of the principal axes solution to the cos θ matrix as outlined by Imbrie (1963) and Klovan (1966). In the original data matrix a series of 7 samples were included from Breidamerkursandur S.E. Iceland, an area at present undergoing deglacierization and a sample of a solifluction deposit generated over coarse sandstone in S. Scotland. These samples were included to investigate the possibility that glacial action *per-se* imparted a distinctive particle-size distribution to a sediment and to investigate the *apparent* similarity (apparent in a field inspection) between the sediment described by Ragg & Bibby (1966) as typical of the hillslope deposits on Broad Law and the deposits on the hillslopes of Upper Weardale. It should be noted that in other analyses of these data there is no significant change in the nature of the dominant factors generated when these extra-regional samples are included or omitted (Falconer 1970).

Tentative classifications were assigned to all samples in the field. It is the basic premise of the present paper that this is not necessarily

157

TABLE 1
PARTICLE-SIZE DATA GROUPED INTO CATEGORIES [1]

PERCENTAGE IN EACH SIZE CATEGORY

GRID REFERENCE	DEPTH	LARGER THAN -3.0φ	-3.0/-1.5φ	-1.5/0.0φ	0.0/1.5φ	1.5/3.0φ	3.0/4.5φ	4.5/6.0φ	6.0/7.5φ	7.5/9.0φ	SMALLER THAN 9.0φ
805435	100	00.0	00.0	05.5	04.5	2.95	32.5	05.6	02.4	04.9	15.1
	125	01.8	00.3	02.9	01.6	01.5	09.3	26.4	13.2	09.0	34.0
	150	03.7	00.5	00.3	00.2	00.8	04.0	32.3	14.2	09.7	34.3
	175	05.5	12.0	09.6	02.9	04.8	15.8	19.1	07.7	05.0	17.6
	200	07.3	17.5	17.0	05.4	08.7	11.9	15.2	05.0	01.6	10.4
	225	49.0	27.6	12.0	02.4	07.2	00.3	00.3	00.3	00.3	00.6
806434	25	00.0	01.2	06.8	11.5	38.3	11.2	03.3	04.1	05.1	18.5
	200	02.1	17.9	31.5	06.1	14.1	12.3	04.0	02.2	01.5	08.3
821443	150	00.0	02.0	06.9	06.5	30.3	22.9	10.2	04.6	06.8	09.8
	200	08.6	06.4	09.3	05.7	17.9	08.9	09.2	08.3	08.7	17.0
823437	75	05.5	01.5	06.3	03.7	18.8	21.3	10.8	05.9	05.6	20.6
	100	01.4	02.6	07.5	03.2	06.1	09.2	22.3	18.7	10.9	18.1
	125	04.8	10.9	17.3	05.0	05.6	07.8	19.4	12.6	07.8	08.8
	140	01.6	09.2	12.4	02.5	04.3	13.0	12.7	13.5	06.9	23.9
	160	08.3	15.4	16.3	05.8	06.2	10.4	11.6	08.7	08.7	08.6
	185	10.0	24.8	19.5	05.3	06.5	06.4	09.5	10.0	05.2	02.8
902394	100	00.0	00.0	12.0	04.3	13.3	12.8	12.4	10.2	09.3	25.7
985352	300	00.0	00.0	17.8	04.2	11.2	11.6	10.8	10.0	08.1	26.3
985381	45	08.9	01.4	02.7	02.2	13.2	15.9	11.7	13.9	09.4	20.7
	80	00.0	01.9	21.9	08.8	07.5	03.3	14.0	19.4	11.2	12.0
	110	26.6	07.7	10.5	05.2	05.8	17.2	09.8	04.9	04.7	07.6
	140	03.6	11.0	12.5	04.6	08.9	26.7	10.1	05.0	05.9	11.7
985392	75	11.8	07.6	06.6	04.8	14.2	16.0	06.7	10.1	08.6	13.6
	135	07.6	07.7	10.1	04.6	07.1	04.7	10.3	14.6	12.1	21.2
	210	42.4	14.4	12.9	04.2	11.0	09.5	02.3	02.1	01.0	00.2
003367	45	02.3	07.5	08.0	05.4	10.7	09.5	08.4	09.9	09.1	29.2
	90	06.6	04.4	05.0	03.0	09.9	15.1	10.3	09.3	08.7	27.7
	135	02.2	00.8	01.8	00.8	04.6	31.8	24.1	13.9	06.1	13.9
	180	04.5	04.7	07.8	04.7	11.3	21.8	16.0	13.4	11.1	13.7
	225	05.6	06.5	07.9	03.1	09.0	14.0	06.9	09.4	09.4	28.2
054383	150	09.8	06.0	07.9	05.5	15.9	12.9	1.02	08.5	14.7	18.6
067384	150	17.3	21.3	29.4	10.6	03.4	02.7	01.7	02.3	00.9	10.4

158

GRID REFERENCE	DEPTH	LARGER THAN -3.0φ	-3.0/-1.5φ	-1.5/0.0φ	0.0/1.5φ	1.5/3.0φ	3.0/4.5φ	4.5/6.0φ	6.0/7.5φ	7.5/9.0φ	SMALLER THAN 9.0φ
068377	120	09.4	04.6	06.0	04.9	11.8	08.8	10.2	09.5	08.8	26.0
068384	75	07.0	05.3	11.7	10.2	23.8	09.6	06.4	04.5	04.7	16.8
	120	00.0	05.1	03.1	03.2	22.3	12.7	06.9	08.7	08.4	29.6
	210	00.0	00.5	16.3	29.2	36.0	05.0	05.0	04.9	02.0	01.1
074345	50	00.0	00.0	08.0	08.5	22.4	08.9	18.1	07.2	06.-	20.4
	105	00.0	00.0	10.5	08.5	19.0	16.5	07.7	06.6	07.2	24.0
173358	150	01.1	01.8	03.4	03.7	11.8	16.2	10.0	10.1	09.2	32.7
205394	90	04.1	00.2	02.7	07.3	36.4	15.4	03.8	03.9	05.4	20.8
	110	02.6	02.3	04.8	04.4	09.9	10.9	21.9	13.2	12.1	17.9
	150	00.0	00.0	14.8	04.2	06.8	14.2	09.2	11.8	09.2	29.8
236363	90	00.0	00.0	00.1	00.4	03.1	04.8	15.2	20.5	15.9	40.0
244335	75	10.2	04.4	03.7	03.7	12.0	21.1	10.0	08.7	06.2	20.0
	135	00.8	01.9	03.6	03.2	13.9	20.6	10.6	11.5	11.0	22.9
S.Scotland	1	00.0	00.0	32.1	04.7	13.5	24.3	11.1	08.1	01.2	05.0
Iceland 1	2	00.0	00.0	10.0	39.0	36.7	11.6	02.7	00.0	00.0	00.0
2	3	00.0	00.0	08.4	14.6	74.0	01.5	00.7	00.0	00.0	00.0
3	4	00.0	00.0	01.1	01.9	46.2	43.0	03.7	02.5	01.1	00.0
4	5	00.0	00.0	00.0	00.1	01.1	47.1	27.2	11.1	08.7	04.8
5	6	02.4	17.6	31.0	16.2	19.2	02.9	00.3	00.4	00.0	00.0
6	7	05.7	06.5	17.8	09.1	20.0	25.9	09.5	05.5	00.0	00.0
7	8	00.0	12.0	06.3	27.1	41.0	07.2	03.2	02.3	00.3	00.0
808370	22	0.5	0.9	4.0	6.1	39.0	31.9	4.9	4.2	3.4	5.6
	30	5.2	4.4	8.6	7.8	29.3	23.9	5.1	3.5	4.0	**8.2**
	90	0.0	1.0	3.4	2.3	3.5	7.6	22.2	26.1	15.1	18.8
825428	180	1.6	2.6	4.9	4.3	26.4	33.6	8.2	5.9	5.1	7.4
	90	5.4	10.2	13.2	5.1	20.1	9.6	6.1	8.4	6.9	15.0
	110	9.4	12.7	9.2	4.7	35.2	12.5	3.3	2.8	3.8	6.4
	160	10.4	12.9	9.1	4.8	25.6	12.3	6.1	3.9	4.9	10.0
826413	45	13.4	12.9	12.9	4.4	8.4	8.7	4.7	7.7	10.5	16.4
	70	7.6	11.6	19.8	9.1	10.3	11.6	4.2	3.3	9.1	13.4
	225	0.8	2.9	10.2	8.2	12.1	11.8	10.7	14.4	12.9	16.0

159

GRID REFERENCE	DEPTH	LARGER THAN -3.0φ	-3.0/-1.5φ	-1.5/0.0φ	0.0/1.5φ	1.5/3.0φ	3.0/4.5φ	4.5/6.0φ	6.0/7.5φ	7.5/9.0φ	SMALLER THAN 9.0φ
836419	22	5.8	16.5	15.1	4.3	5.8	30.5	20.2	0.4	1.0	0.4
	45	4.8	4.1	2.9	2.2	18.0	25.3	17.3	11.4	9.7	4.3
	60	2.7	1.6	1.9	1.9	8.9	9.2	19.7	17.2	11.3	25.6
840397	30	4.2	11.8	6.0	3.8	15.1	16.7	10.2	8.2	8.8	15.2
	60	8.7	5.9	6.1	4.5	13.9	19.5	18.7	9.6	7.1	16.0
	90	8.8	5.5	4.5	3.2	13.1	17.1	10.2	8.2	7.7	21.7
	450	4.9	2.7	1.4	2.2	6.5	10.7	16.8	20.4	18.5	20.9
852407	180	5.1	10.9	9.2	2.8	18.0	21.1	10.9	5.5	6.2	10.3
	150	6.0	8.1	7.7	3.7	15.5	17.2	15.1	8.7	7.0	11.0
	105	11.0	9.0	9.7	4.6	14.1	19.1	10.3	6.8	6.0	9.4
	90	2.6	6.7	12.3	6.1	9.9	7.6	13.9	13.6	10.3	17.0
	60	9.4	11.9	12.7	4.0	16.3	15.8	6.2	6.5	5.3	10.9
	30	3.8	7.3	7.3	11.8	27.6	29.2	8.3	3.3	1.0	0.4
865410	195	4.1	0.8	2.3	4.9	11.1	35.2	14.4	5.5	7.2	14.5
	120	12.3	15.5	19.2	8.9	8.9	9.8	6.1	8.2	5.9	5.2
	45	0.0	1.8	1.3	4.8	12.7	12.6	9.1	13.1	13.0	13.6
	22	1.4	1.2	7.7	6.2	13.4	16.1	9.8	12.2	11.5	20.5
862353	180	7.2	9.8	7.8	3.4	10.2	26.2	18.2	7.2	5.7	4.3
	120	5.0	9.9	9.3	5.3	7.8	17.2	24.1	6.9	8.9	5.6
	105	1.4	0.6	2.5	1.9	5.2	22.4	22.0	11.8	10.8	21.4
	75	0.0	0.0	10.4	7.8	12.4	27.3	18.1	8.7	8.7	6.6
869360	150	6.3	4.7	8.1	2.6	9.2	16.9	14.6	9.9	7.7	20.0
	75	11.4	26.6	18.7	7.1	5.1	8.1	7.1	4.1	1.2	4.6
	60	5.3	10.2	16.9	11.0	13.4	14.5	5.9	8.2	7.4	7.2
869394	90	31.8	17.8	8.6	3.8	9.7	5.1	4.6	5.3	4.1	9.2
	60	7.4	5.4	5.9	7.2	16.9	14.5	6.7	7.4	6.5	22.1
	30	4.2	4.8	10.7	7.1	22.4	10.8	8.1	9.7	10.1	12.1
873379	60	2.1	5.7	11.9	6.5	17.8	14.4	7.5	6.8	7.1	10.2
	180	7.0	4.6	3.4	2.1	16.9	15.7	7.5	7.1	8.1	27.0
883346	360	6.6	12.9	17.9	7.2	13.6	7.6	6.8	8.9	7.3	11.2
	90	1.4	5.7	12.1	7.2	14.3	13.3	12.0	4.2	13.5	16.3
	30	0.8	1.9	7.5	7.4	23.2	25.3	6.6	6.5	6.8	14.0

GRID REFERENCE	DEPTH	LARGER THAN -3.0φ	-3.0/ -1.5φ	-1.5/ 0.0φ	0.0/ 1.5φ	1.5/ 3.0φ	3.0/ 4.5φ	4.5/ 6.0φ	6.0/ 7.5φ	7.5/ 9.0φ	SMALLER THAN 9.0φ
888413	195	1.2	3.1	6.4	4.6	11.6	29.8	14.8	7.5	6.5	14.5
	165	4.4	11.9	11.9	7.6	14.2	15.7	9.3	6.5	5.5	14.0
903331	90	2.8	5.1	8.1	5.7	18.3	20.8	11.8	6.4	5.0	16.0
	105	5.4	7.4	13.2	12.7	30.9	8.3	3.8	2.8	6.8	8.7
	90	2.3	6.3	18.4	16.8	25.6	15.9	4.1	4.7	3.6	3.8
912348	90	7.8	7.2	7.6	4.6	35.8	34.6	5.4	2.8	2.3	1.9
	22	10.9	11.1	9.7	4.7	21.6	29.0	5.9	2.4	3.2	1.5
927440	45	8.6	33.2	32.2	7.4	5.4	5.2	2.0	0.7	0.4	1.5
	30	9.8	26.1	34.3	7.4	3.7	2.7	3.0	3.2	3.6	3.9
931373	180	15.6	22.9	14.3	3.2	14.8	14.3	3.6	4.2	2.8	6.2
	75	25.0	19.8	8.9	4.2	11.4	14.1	4.6	4.1	2.6	4.3
	60	8.4	6.8	10.9	7.7	15.1	17.3	12.4	7.5	6.2	5.3
947386	90	6.1	8.2	10.5	5.2	16.9	17.7	9.4	8.1	7.2	7.7
	60	4.4	7.8	7.6	7.1	19.3	20.9	11.5	5.7	3.9	10.7
	30	0.6	2.2	3.9	2.5	18.6	23.9	10.7	8.6	9.3	12.0
952440	300	24.8	15.2	11.8	5.6	16.9	10.1	5.1	2.3	2.5	20.7
	240	5.4	11.2	15.2	6.4	20.1	22.2	7.4	3.7	3.4	5.7
	150	11.0	29.3	18.5	5.7	12.9	10.4	4.4	2.5	2.1	5.0
	75	7.3	13.9	9.8	4.3	8.5	9.9	8.4	8.2	6.5	3.7
952449	75	0.0	1.5	7.9	9.9	32.8	21.7	4.2	5.6	5.9	23.3
	45	0.2	3.5	10.1	10.2	13.8	18.9	12.8	15.8	8.9	10.5
	35	0.0	0.0	3.9	5.2	2216	32.6	9.9	7.2	6.8	5.8
	30	0.3	0.2	5.3	6.6	14.8	17.4	9.7	10.7	13.3	11.8
962334	240	5.4	28.7	38.7	10.4	5.7	4.8	2.7	1.4	1.4	21.7
	75	12.6	26.9	21.3	6.4	7.5	5.3	4.9	3.8	3.4	0.8
	30	10.1	22.9	22.3	6.8	5.8	5.1	3.9	3.4	7.3	6.9
	15	4.3	5.7	12.9	11.2	21.5	8.2	8.1	8.3	8.9	12.4
962450	30	1.8	3.1	7.7	8.2	27.1	14.4	3.9	6.8	7.4	10.9
	60	3.5	4.1	3.9	10.5	54.4	13.1	4.3	1.8	2.2	13.6
992403	150	2.9	9.1	10.4	5.7	24.3	17.9	7.5	6.0	5.4	2.2
	120	5.3	5.4	6.5	6.2	11.7	12.7	10.1	7.9	9.6	10.8
	90	1.4	4.5	5.9	4.7	17.7	16.8	14.1	8.1	7.8	19.0

GRID REFERENCE	DEPTH	LARGER THAN -3.0φ	-3.0/ -1.5φ	-1.5/ 0.0φ	0.0/ 1.5φ	1.5/ 3.0φ	3.0/ 4.5φ	4.5/ 6.0φ	6.0/ 7.5φ	7.5/ 9.0φ	SMALLER THAN 9.0φ
997324	90	11.6	18.5	16.2	11.5	14.4	11.3	6.3	3.6	3.4	5.2
	45	2.6	1.8	7.7	11.1	27.1	15.7	6.6	7.2	5.6	14.6
	30	0.1	2.0	8.0	10.6	27.5	18.1	6.6	5.9	7.5	12.7
003348	120	18.7	19.6	16.4	7.3	10.4	12.5	5.1	3.8	2.7	3.5
	90	6.7	7.3	24.3	23.7	21.4	6.6	2.4	3.0	2.9	1.7
	15	17.4	21.2	19.5	9.7	18.3	4.6	2.1	2.5	3.4	1.3
010423	150	9.3	2.8	33.9	18.3	18.5	6.4	1.8	2.4	2.8	3.8
	60	2.7	11.8	20.7	18.6	33.0	4.9	1.5	0.7	1.4	4.7
	22	7.4	7.4	15.2	16.3	31.3	6.9	1.5	3.8	3.4	6.8
049341	120	9.2	5.7	6.7	4.6	21.1	28.3	4.3	4.9	4.4	10.8
	60	18.7	13.4	14.8	13.4	21.7	4.1	1.5	1.5	2.5	8.4
	45	10.2	14.8	21.1	15.6	27.1	3.2	2.9	1.9	2.3	0.9
	22	19.7	23.1	19.1	12.7	18.4	2.6	1.3	1.8	1.0	0.3
074345	270	11.3	10.6	12.9	11.2	16.3	14.2	5.3	5.5	5.8	7.4
080362	60	1.2	3.6	13.3	47.1	25.3	6.4	2.9	1.0	0.8	0.4
	38	0.0	2.1	29.8	25.9	20.3	6.6	1.4	2.9	2.8	8.2
	22	3.1	3.3	20.8	17.3	22.3	9.9	3.7	3.5	5.9	8.7
097408	45	8.9	10.2	9.1	9.7	14.7	12.7	8.2	6.9	5.9	13.7
	15	4.2	5.9	7.7	13.0	28.8	12.4	5.5	7.9	7.8	6.8
095353	45	10.1	6.1	9.9	8.9	7.2	10.7	1.3	8.5	6.9	10.4
	22	3.2	5.9	28.0	23.1	26.9	6.7	1.2	2.2	2.2	0.6
097493	60	9.6	11.1	16.6	10.7	18.4	14.8	7.0	3.8	4.6	3.4
	270	4.0	8.1	12.1	6.2	16.2	26.2	11.6	3.9	6.3	5.4

[1]Samples are identified by 6 digit grid references to British Topographic Series grid zone NY, NZ.

TABLE 2

VARIMAX FACTOR MATRIX - ALL DATA[1]

Sample[2] No.	Field Classification	Communality	Factor 1	Factor 2	Factor 3	Factor 4	Factor 5
P1	S	0.9628	0.4571	0.0638	0.6113	0.6008	0.1231
P2	T	0.9529	0.9612	0.0828	-0.0295	0.1407	0.0378
P3	T	0.9004	0.9389	0.0706	-0.0887	0.0767	0.0122
P4	T	0.9620	0.7476	0.4649	0.0654	0.4261	0.0342
P5	R	0.9510	0.4935	0.7248	0.2071	0.3630	0.0863
P6	R	0.9546	-0.0014	0.8219	-0.0008	0.0418	0.5266
P7	S	0.9900	0.4716	0.1248	0.8402	0.1715	0.1292
P8	R	0.9071	0.2699	0.7466	0.4020	0.2392	-0.2408
P9	S	0.9897	0.4579	0.1435	0.6843	0.5371	0.0515
P10	T	0.9920	0.7138	0.4295	0.4851	0.2085	0.1390
P11	T	0.9782	0.7108	0.1939	0.4505	0.4599	0.1442
P12	T ?	0.9302	0.8828	0.2220	0.0909	0.2575	-0.1643
P13	T	0.9381	0.6325	0.6079	0.1221	0.2921	-0.2613
P14	T	0.9743	0.8623	0.3930	0.1036	0.2354	-0.1002
P15	R	0.9872	0.5442	0.7434	0.1706	0.3144	-0.1017
P16	R	0.9445	0.3017	0.8834	0.1264	0.2229	-0.0859
P17	T	0.9793	0.8705	0.2040	0.3586	0.2107	-0.0831
P18	T	0.9419	0.8375	0.2850	0.3328	0.1520	-0.1595
P19	S?	0.9821	0.8397	0.1968	0.2775	0.3527	0.1923
P20	T	0.9125	0.6882	0.4391	0.2683	0.0688	-0.4114
P21	R	0.8716	0.3950	0.6471	0.1000	0.4378	0.3085
P22	R	0.9609	0.5363	0.4453	0.2489	0.6407	-0.0502
P23	S	0.9788	0.6517	0.4481	0.3718	0.4046	0.2268
P24	S	0.9826	0.8645	0.4491	0.1768	0.0427	0.0216
P25	R	0.8754	0.0646	0.7549	0.1305	0.2182	0.4867
P26	S?	0.9685	0.8874	0.2960	0.2888	0.0839	0.0540
P27	S?	0.9803	0.8853	0.2412	0.2277	0.2483	0.1574
P28	T?	0.9833	0.6826	0.0782	0.0281	0.7124	-0.0554
P29	S?	0.9579	0.7918	0.3292	0.2760	0.3789	-0.0528
P30	T	0.9563	0.8701	0.3079	0.2318	0.1884	0.1236

[1] The sequence of samples in this table corresponds exactly with -TABLE 1
[2] Sample numbers relate to initial sampling P- Purposive   Sample
R- Random Sample.

P31	T	0.9886	0.7273	0.4031	0.4174	0.3034	0.1754
P32	R	0.9651	0.2346	0.9379	0.1658	−0.0184	−0.0511
P33	T	0.9938	0.8716	0.3197	0.2897	0.1176	0.1846
P34	R	0.9908	0.5695	0.4044	0.6763	0.1862	0.1044
P35	S	0.9582	0.8057	0.1209	0.4866	0.1656	0.1739
P36	R	0.9475	0.1536	0.2903	0.8925	0.1039	−0.1794
P37	T	0.9402	0.7486	0.1615	0.5512	0.2178	−0.0495
P38	T	0.9652	0.7488	0.1827	0.5452	0.2715	−0.0164
P39	T	0.9634	0.9083	0.0950	0.2740	0.2103	0.0999
P40	S	0.9907	0.5268	0.0844	0.7581	0.2543	0.2582
P41	T	0.9336	0.8688	0.1939	0.1936	0.3138	−0.0733
P42	T?	0.9333	0.8877	0.2305	0.2279	1.1592	−0.1216
P43	Clay	0.9949	0.9954	0.0166	0.0179	−0.0451	0.0376
P44	T?	0.9807	0.7461	0.2770	0.2781	0.4679	0.2258
P45	T?	0.9858	0.8334	0.1085	0.3293	0.4088	0.0621
P46		0.8669	0.3502	0.4416	0.3827	0.5126	−0.3742
P47		0.8739	0.0760	0.2072	0.8821	0.1765	−0.1265
P48		0.9189	0.0501	0.0819	0.9382	0.0916	0.1450
P49		0.9792	0.1580	−0.0035	0.6641	0.6936	0.1794
P50		0.9877	0.4681	0.0166	−0.0356	0.8701	−0.1003
P51		0.9861	0.0339	0.7526	0.5825	0.0526	−0.2765
P52		0.9818	0.2388	0.4581	0.5251	0.6540	−0.1066
P53		0.9042	0.0849	0.2934	0.8846	0.1682	0.0123
R1	R	0.9898	0.2803	0.0641	0.7194	0.6081	0.1408
R2	R	0.9958	0.3619	0.2630	0.6965	0.5410	0.1336
R3	T	0.8684	0.8920	0.1108	0.0051	0.1900	−0.1562
R4	R	0.9909	0.3941	0.1217	0.5479	0.7154	0.0931
R5	T	0.9775	0.6142	0.5011	0.5466	0.2141	0.0677
R6	S	0.9549	0.2603	0.4223	0.7308	0.3214	0.2673
R7	S	0.9771	0.4132	0.5104	0.6063	0.3368	0.2545
R8	S	0.9774	0.6463	0.6762	0.2276	0.1510	0.1669
R9	S	0.9493	0.5179	0.6909	0.3880	0.2179	−0.0761

R10	R	0.9661	0.8017	0.2914	0.3861	0.2621	-0.1439
R11	R	0.9657	0.2695	0.5268	0.0966	0.7679	-0.1288
R12	R	0.9459	0.5238	0.1910	0.3439	0.7172	0.0496
R13	S/T	0.9698	0.9506	0.1132	0.1317	0.1885	0.0194
R14	S	0.9679	0.6869	0.3952	0.3688	0.4377	0.1109
R15	T/S	0.9763	0.7247	0.3255	0.2800	0.5089	0.0883
R16	T	0.9935	0.7939	0.2954	0.3053	0.3702	0.2141
R17	T	0.9339	0.9188	0.1604	0.0809	0.2377	0.0319
R18	T	0.9834	0.5308	0.4194	0.4259	0.5806	0.0853
R19	S	0.9789	0.6429	0.3910	0.3642	0.5270	0.0492
R20	S/T	0.9967	0.5372	0.5113	0.3597	0.5472	0.1340
R21	T	0.9827	0.8187	0.4219	0.2826	0.1836	-0.1440
R22	S	0.9895	0.5096	0.5879	0.4367	0.4208	0.1282
R23	T	0.9915	0.2159	0.2757	0.6368	0.6790	0.0478
R24	R	0.9652	0.5894	0.1017	0.2451	0.7374	0.0603
R25	R	0.9865	0.3643	0.8339	0.2924	0.2655	-0.0489
R26	R	0.9791	0.9278	0.0636	0.2955	0.1352	0.0931
R27	S	0.9859	0.8325	0.1958	0.3864	0.3233	-0.0275
R28	R	0.9853	0.4700	0.4012	0.1889	0.7535	-0.0065
R29	T	0.8973	0.5673	0.4378	0.1375	0.5870	-0.1429
R30	T	0.9898	0.8523	0.0845	0.0676	0.4998	-0.0430
R31	S	0.9830	0.5488	0.2000	0.3349	0.6984	-0.2046
R32	T	0.9949	0.8267	0.3199	0.2074	0.4053	0.0437
R33	S	0.9592	0.2277	0.9167	0.1238	0.2267	-0.0174
R34	T	0.9815	0.4589	0.6180	0.4811	0.3727	-0.1365
R35	S	0.9600	0.3021	0.7745	0.1191	0.1212	0.4899
R36	T	0.9841	0.7477	0.3105	0.4683	0.2690	0.1923
R37	T	0.9720	0.6125	0.3599	0.6227	0.2811	0.0229
R38	S	0.9743	0.5318	0.5128	0.5061	0.3909	0.1398
R39	T	0.9776	0.8097	0.2077	0.3766	0.2630	0.2605
R40	T	0.9884	0.5420	0.6910	0.4221	0.1830	-0.0741
R41	T	0.9513	0.7204	0.3646	0.4352	0.3179	-0.0948

R42	S	0.9836	0.5503	0.1681	0.5975	0.5419	0.0439
R43	T	0.9867	0.6402	0.1763	0.2799	0.6828	−0.0345
R44	T	0.9835	0.6100	0.5221	0.4252	0.3975	−0.0066
R45	T	0.9898	0.6537	0.2739	0.4768	0.5085	0.0391
R46	R	0.9895	0.3434	0.4184	0.8103	0.1891	0.0644
R47	R	0.9946	0.2616	0.4590	0.7538	0.3581	−0.1378
R48	R	0.9913	0.2092	0.2424	0.6282	0.6762	0.1920
R49	R	0.9955	0.2281	0.4380	0.4740	0.7045	0.1748
R50	R	0.9323	0.0912	0.9325	0.1548	0.0881	−0.1508
R51	R	0.9668	0.1848	0.9319	0.1468	0.0229	−0.2051
R52	S	0.9783	0.2303	0.7985	0.3041	0.3811	0.2236
R53	S	0.9904	0.2469	0.7701	0.1909	0.3652	0.4080
R54	T?/S	0.9826	0.5350	0.4819	0.4335	0.5254	0.0085
R55	S	0.9957	0.5827	0.4451	0.4627	0.4916	0.0480
R56	R	0.9897	0.5551	0.3438	0.5056	0.5508	0.0660
R57	R/S?	0.9926	0.7298	0.1635	0.3970	0.4946	0.1762
R58	R	0.9753	0.2453	0.7517	0.3548	0.2790	0.3824
R59	R	0.9900	0.3322	0.5129	0.5291	0.5802	−0.0054
R60	R	0.9300	0.1733	0.8651	0.2684	0.2728	0.0717
R61	S	0.9663	0.7789	0.5263	0.2116	0.1559	0.1166
R62	R	0.9942	0.4107	0.1518	0.7738	0.4468	0.0631
R63	R	0.9301	0.5808	0.2968	0.4223	0.5293	−0.2149
R64	S/T	0.9904	0.5156	0.0638	0.4995	0.6843	0.0516
R65	S	0.9801	0.8287	0.1196	0.4098	0.3334	0.0028
R66	R	0.9598	0.0628	0.9019	0.2102	0.0815	−0.3025
R67	S	0.9704	0.2520	0.9271	0.1781	0.1246	0.0122
R68	S	0.9609	0.3940	0.8755	0.1842	0.0599	−0.0412
R69	S	0.9826	0.5566	0.4385	0.6589	0.2096	−0.0491
R70	S/T	0.9922	0.5710	0.2240	0.7004	0.3491	0.0590
R71	R	0.9456	0.1473	0.1289	0.8836	0.2905	0.2051
R72	R	0.9805	0.4905	0.3749	0.6209	0.4579	0.0646
R73	T	0.9844	0.8561	0.2968	0.3249	0.2191	0.0988
R74	T	0.9890	0.7656	0.2127	0.4346	0.4093	0.0362

R75	R	0.9873	0.2867	0.7915	0.4280	0.3070	0.0358
R76	S	0.9990	0.5520	0.2147	0.7285	0.3364	0.0660
R77	R	0.9979	0.5170	0.1875	0.7318	0.3994	0.0215
R78	R	0.9969	0.2237	0.8608	0.2556	0.3372	0.1638
R79	R	0.9614	0.1477	0.6117	0.7098	0.1166	-0.2192
R80	S	0.9919	0.1146	0.8612	0.4465	0.1396	0.1350
R81	R	0.8965	0.1724	0.6486	0.6068	0.0833	-0.2665
R82	R	0.9911	0.1532	0.5158	0.8283	0.0963	-0.0791
R83	R	0.9960	0.2555	0.4668	0.8313	0.1390	0.0486
R84	R	0.9738	0.4371	0.2991	0.4996	0.6294	0.2182
R85	R	0.9833	0.2476	0.7156	0.5900	0.0699	0.2385
R86	R	0.9870	0.1085	0.6864	0.7022	0.1041	-0.0135
R87	R	0.9888	0.0510	0.8716	0.4407	0.0871	0.1571
R88	S	0.9855	0.4064	0.6252	0.5323	0.3725	0.0858
R89	R	0.7250	0.0719	0.3284	0.7391	0.0526	-0.2509
R90	R	0.9429	0.2456	0.4972	0.7087	0.0387	-0.3627
R91	R	0.9796	0.3588	0.4846	0.7411	0.1811	-0.1843
R92	T	0.9867	0.6117	0.5348	0.4599	0.3215	0.1078
R93	S	0.9689	0.4015	0.3203	0.7745	0.3177	0.0659
R94	R	0.8889	0.5804	0.5940	0.3602	0.2499	0.0845
R95	R	0.9839	0.1047	0.5449	0.7730	0.1142	-0.2558
R96	S/T	0.9954	0.2979	0.6486	0.5550	0.4217	-0.0127
R97	R	0.9949	0.4161	0.4000	0.4247	0.6921	-0.0485

a valid procedure. Thus some sediments which were only tentatively identified are accompanied by a "?". Investigation of the field notes and site descriptions reveals that the group of samples for which factor 1 is dominant seems to be the group of samples described as till. There are differences between the field classification of the sample as "till" and the samples which load predominantly on Factor 1. Whether this difference is attributable to incorrect field classification or reflects some deficiency in the technique employed here cannot be readily ascertained. However, a consideration of all the samples loading predominantly on Factor 1 shows that many of their characteristics are very similar to the characteristics of "tills" identified by other investigators working in adjacent regions. Figure 6 illustrates the comparision of the texture of deposits considered to be till by Vincent (1969) and Beaumont (1967). Brief analyses of the content of larger particles in these samples indicated that striated stones were common (25%+) and that the lithology was very similar to that noted by Atkinson (1968) for his till samples. Similarities between the properties of the samples which loaded predominantly on factor 1 and the properties of the East Durham Till (Beaumont 1967) are noted in Table 3.

Comparison of the sand silt and clay composition of the factor 1 group with the deposits of Northern England generally considered to be till show the similarities noted in Table 4.

On the basis of the comparisons noted above factor 1 was considered to represent till characteristics in the data. Thus an objective measure of the influence of till creating process is available as the factor 1 loading for each sample. Inspection of table 2 indicates that many samples have secondary loadings on other factors and in many cases the secondary loadings are very similar to the primary loadings. It is thus necessary to investigate the nature of the other factors.

There is very little published data available for the hillslope deposits of the northern Pennines. Factor 2 was a group of deposits which had field descriptions indicating that they were soil-creep and water-washed hillslope deposits. It was almost impossible to verify this indication quantitatively in the regional context. However, the published particle-size curve for rainwashed slope detritus (Twenhofel, 1932, p. 237) displays characteristics which can be identified for this group of deposits. Assessing the "average" distribution curve for the factor

TABLE 3

	E. Durham Till	Factor 1 Group
Median diameter (phi units)	4.0 - 6.5	4.22 (range 3.2-6.6)
Mean diameter (phi units)	4.0 - 6.5	4.29 (3.6 - 6.6)
Sorting	3.1 - 6.4	4.51 (2.7 - 5.8)

TABLE 4

	% Sand	% Silt	% Clay
Leeds Till	33	46	21
Hessle Till	23	51	26
Purple Till	20	40	25
E. Durham Till	30	40	30
Upper Weardale Factor 1 Group	38	37	25

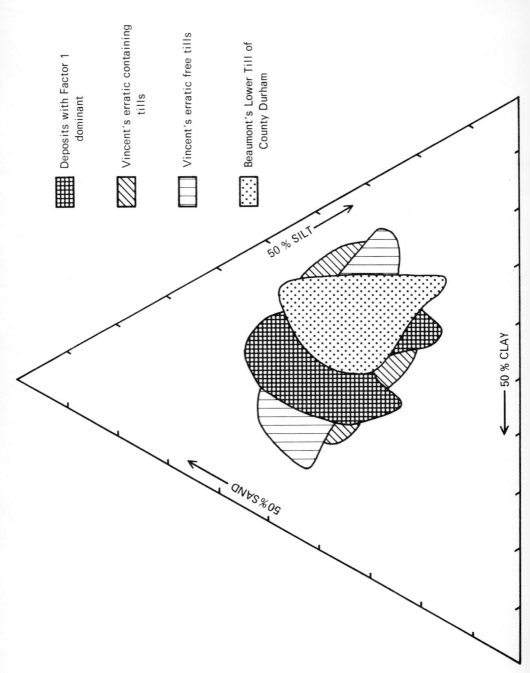

Fig. 6 Textural comparison of Tills and the factor 1 grouping of samples.

2 group of deposits, it is apparent that Twenhofel's specimen deposit is similar. Indeed the photograph and description published by Twenhofel (1932 p. 236) are also similar to the appearance and field description of the Weardale deposits which load highly on factor 2.

Samples which load predominantly on factor 3 approximate more closely to a log-normal distribution than any other group emerging from this analysis. This indicates that these deposits are better sorted than any other group. Factor 3 deposits also have the charactertistics of low clay content and sandy texture the field notes indicate that these deposits in many cases are disintegrating bedrock. It would appear that the better sorting of this group of sandy sediments is a characteristic associated with the sandstone bedrock and thus inherited by the deposits which were sampled in this study. This factor 3 is associated with decomposing bedrock.

Factor 4 is a factor which has loadings of predominantly silty material. Identification of this group is difficult. The field notes indicate a more disturbed deposit than the solifluction/hillslope deposits of the type loading on factor 2. The deposits loading dominantly on factor 4 are conspicuously more silty than those loading on factor 2. In Washburn's (1967) study of the periglacial deposits of Greenland a similar texture is noted and Washburn emphasises the importance of the action of frost in silty materials. Atkinson's description of fossil ice-wedges and polygons in some slope deposits similar in description to this group of deposits loading predominantly on factor 4 tends to emphasise the similarities and factor 4 is thus tentatively associated with the process which Wasburn (1967) terms gelifluction.

No group of samples load predominantly on factor 5. The factor score matrix (Table 5) indicates that the factor 5 is associated with high values of gravel content, high values for clay content and high values of fine sand/coarse silt content. Further examination indicates that factor 5 is a compensatory factor which greatly improves the explanation of total variance by accommodating samples which have extreme values in these portions of the particle-size curve. It is thus regarded as a significant contribution to the solution even although no samples have a dominant loading in this category.

Having thus tentatively associated a genetic significance with each factor it is now possible to interpret the interaction of the factors as interaction of geomorphic processes. Thus the magnitude of the factor loadings can be taken as the measure of the contribution of each process to the creation of the deposit.

TABLE 5

VARIMAX FACTOR SCORE MATRIX (TOTAL DATA)

Size Categories in phi units	FACTOR 1	FACTOR 2	FACTOR 3	FACTOR 4	FACTOR 5
<-3.0	-0.0204	1.4703	-0.3667	0.0604	2.0516
-3.01-1.5	-0.1386	1.9884	-0.3781	0.1647	0.4421
-1.5/0.0	0.0821	1.8751	0.4720	-0.1255	-1.6855
0.0/1.5	-0.0043	0.5036	1.2906	-0.3744	-0.9051
1.5/3.0	0.1354	-0.0481	2.7442	0.3108	0.8467
3.0/4.5	0.4026	-0.0533	0.1327	2.8174	0.0898
4.5/6.0	1.1492	0.1469	-0.5450	0.9792	-0.7709
6.0/7.5	1.1764	0.1238	-0.1755	0.0191	-0.3916
7.5/9.0	0.9756	0.0714	0.0089	-0.0560	-0.1217
79.0	2.4832	-0.0861	0.1152	-0.8915	0.6349

This means that there now is opportunity to examine the interactions of processes acting to create the suite of deposits at present existing in Upper Weardale. Examination of this interaction can be by a series of bivariate plots of factor loadings as proposed by Klovan (1966) or by stratigraphic diagrams indicating the relative important of each process. It is surprising to find that there is an apparent continuum of deposit types and the groups identified by dominant loading are not the isolated groups which would be expected in an area where environments of deposition are clearly defined. Thus unlike Klovan's (1966) results, the results presented in figure 7, illustrate the apparently arbitrary nature of the boundaries of the regions of factor dominance for a series of bivariate plots of the samples. The plots are for the random sample specimens only to avoid congestion of the diagram. The samples are identified by their field classification.

In examining the stratigraphic record of sample sites it was possible to consider only the dominant factor. Thus the layers of surficial deposit were "identified" by a single factor label as till (factor 1), rainwashed slope material (factor 2), decomposing bedrock (factor 3), gelifluction material (factor 4). Figure 8 illustrates this for a particular site in Upper Weardale with the photograph of the stratigraphy paralleled by field notes and by a stratigraphic column representing the dominant factor for each layer. This however represents the use of factor analysis of the particle-size data to simply eliminate an in-the-field-assessment of the deposits and substitute a dominant factor decision. The real value of the factor analysis results is the relative importance of the several factors in explaining the true nature of the deposits.

Fig. 7 illustrates the continuum existing between each pair of factors. Thus any single specimen may be the product of interaction of any two factors. This initially seems to offer some difficulties of logic. Whilst the existence of soliflucted till could be demonstrated by loadings on factors 1 and either factor 2 or factor 4 it is strange that some samples with dominant loadings on factor 3 could have a large secondary loading on factor 1. The implication of this is that decomposing bedrock has been influenced by glacial action. Fig. 9 shows a series of stratigraphic diagrams for some of the sites sampled in this study. The interaction of the various factor influences can be seen in this diagram and the interpretation of this interaction is more valuable than the single process definition based on dominant factor classification. The peculiar combination of the influence of factors 1 and 3 noted above can be explained by a model of glacial action proposed by Vincent (1969). This is presented as Fig. 10 and is expanded by the addition of a diagram of the resulting debris distribution.

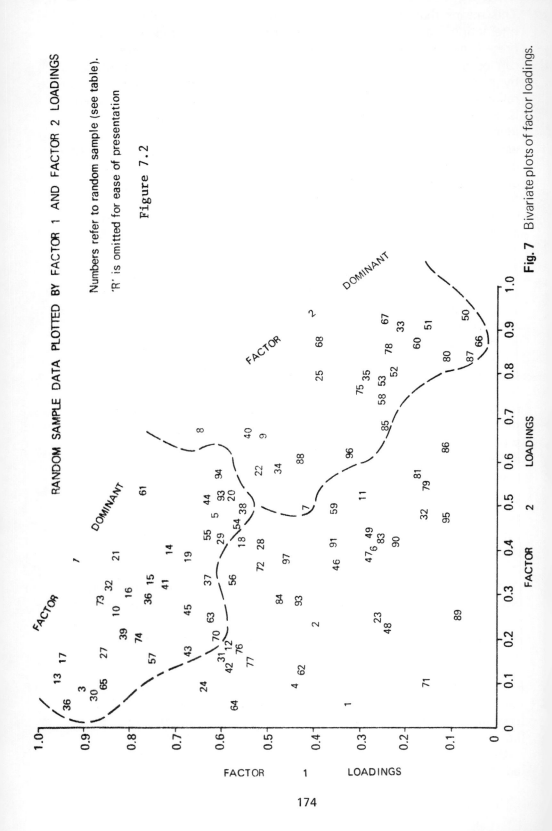

RANDOM SAMPLE DATA PLOTTED BY FACTOR 1 AND FACTOR 2 LOADINGS

Numbers refer to random sample (see table).

'R' is omitted for ease of presentation

Figure 7.2

Fig. 7 Bivariate plots of factor loadings.

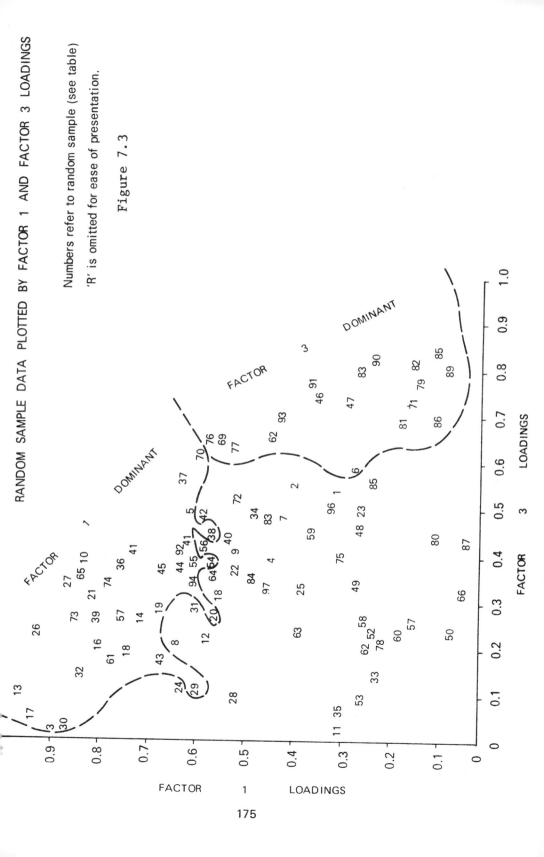

RANDOM SAMPLE DATA PLOTTED BY FACTOR 1 AND FACTOR 3 LOADINGS

Numbers refer to random sample (see table)

'R' is omitted for ease of presentation.

Figure 7.3

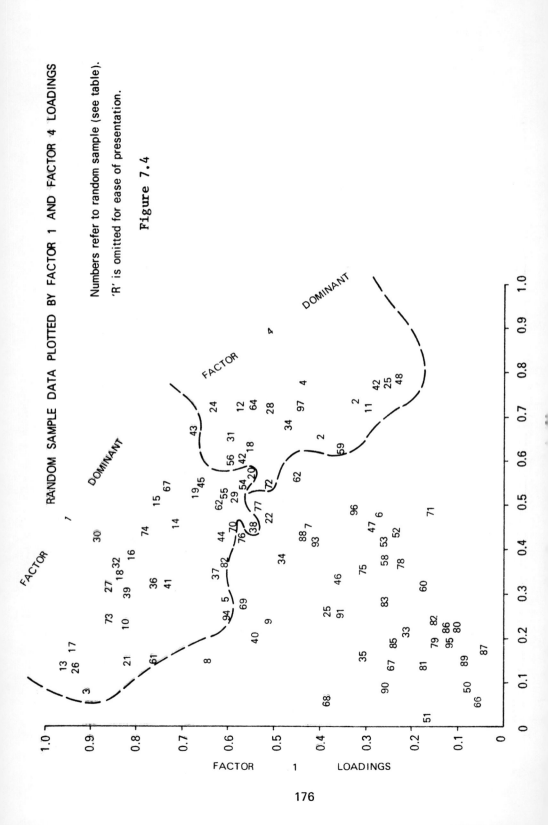

RANDOM SAMPLE DATA PLOTTED BY FACTOR 1 AND FACTOR 4 LOADINGS

Numbers refer to random sample (see table).

'R' is omitted for ease of presentation.

Figure 7.4

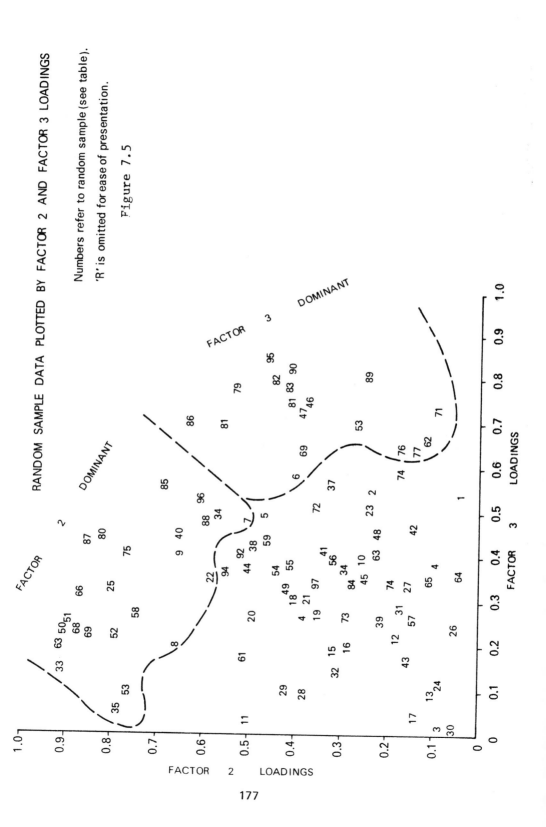

RANDOM SAMPLE DATA PLOTTED BY FACTOR 2 AND FACTOR 3 LOADINGS

Numbers refer to random sample (see table).
'R' is omitted for ease of presentation.

Figure 7.5

177

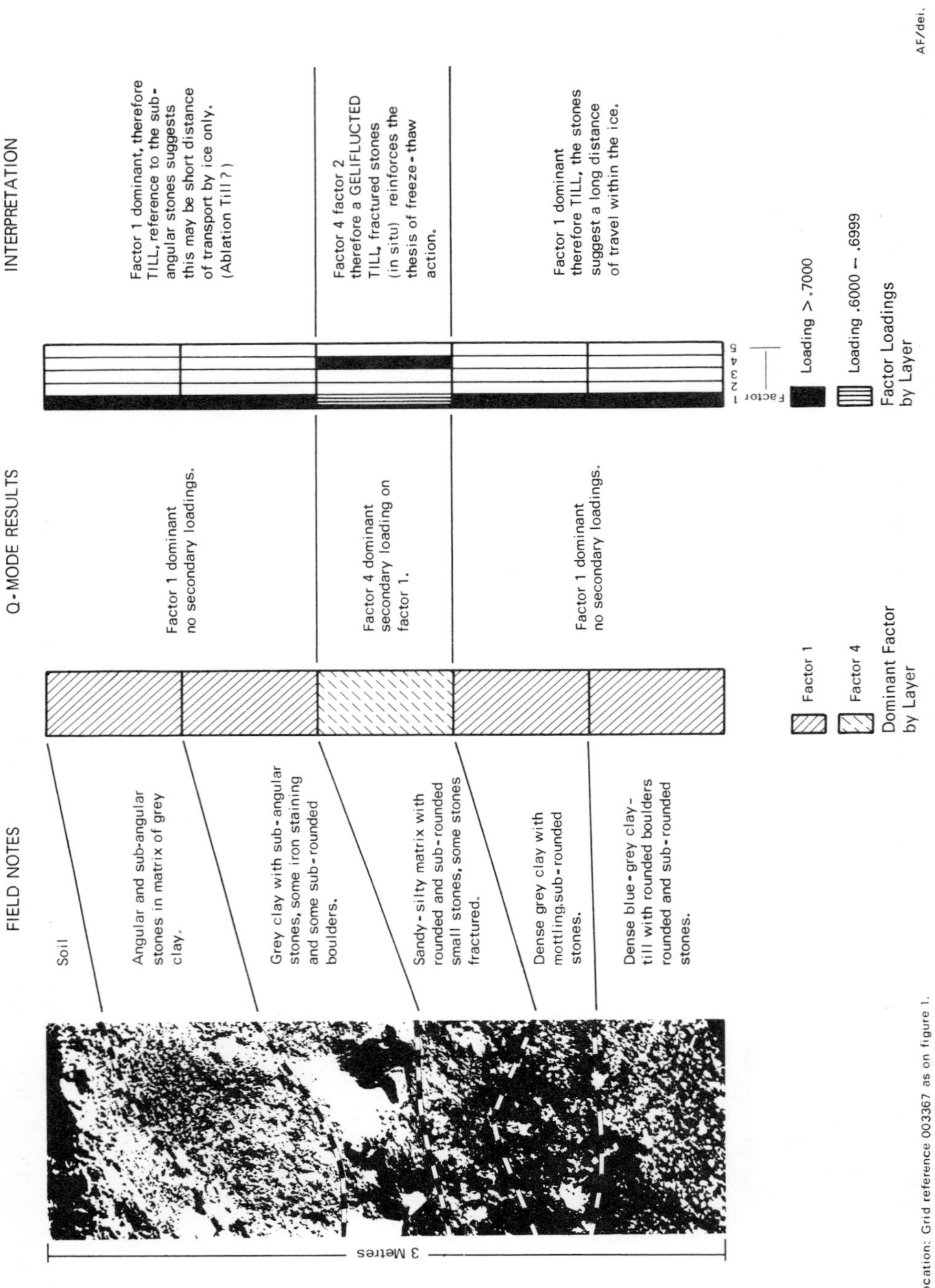

Fig. 8 Analysis of a section in the deposits of Upper Weardale.

INTERPRETATION

Factor 1 dominant, therefore TILL, reference to the sub-angular stones suggests this may be short distance of transport by ice only. (Ablation Till?)

Factor 4 factor 2 therefore a GELIFLUCTED TILL, fractured stones (in situ) reinforces the thesis of freeze-thaw action.

Factor 1 dominant therefore TILL, the stones suggest a long distance of travel within the ice.

Loading > .7000

Loading .6000 ~ .6999

Factor Loadings by Layer

Q-MODE RESULTS

Factor 1 dominant no secondary loadings.

Factor 4 dominant secondary loading on factor 1.

Factor 1 dominant no secondary loadings.

Factor 1

Factor 4

Dominant Factor by Layer

FIELD NOTES

Soil

Angular and sub-angular stones in matrix of grey clay.

Grey clay with sub-angular stones, some iron staining and some sub-rounded boulders.

Sandy-silty matrix with rounded and sub-rounded small stones, some stones fractured.

Dense grey clay with mottling, sub-rounded stones.

Dense blue-grey clay-till with rounded boulders rounded and sub-rounded stones.

3 Metres

Site Location: Grid reference 003367 as on figure 1.

AF/dei.

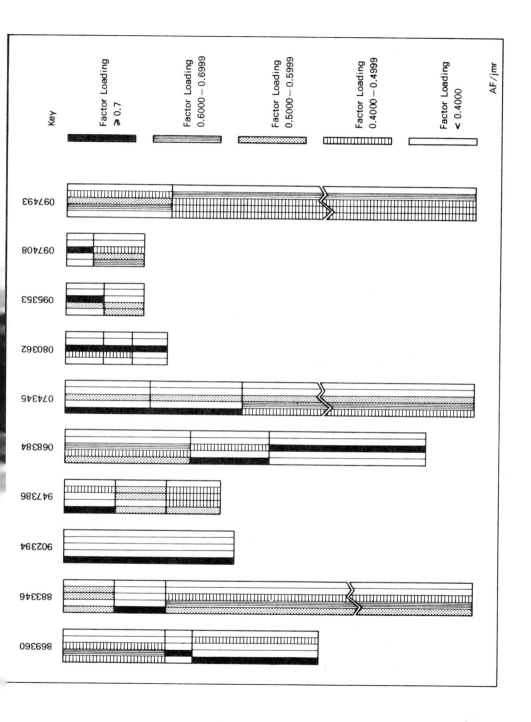

Fig. 9 Some stratigraphic diagrams of factor loadings for sites in Upper Weardale.

VINCENT'S MODEL OF STRONG RELIEF CONTROL ON ICE MOVEMENT AND THE RESULTING DEBRIS DISTRIBUTION

ICE ACTION AT MAXIMUM GLACIATION
(after Vincent 1969)

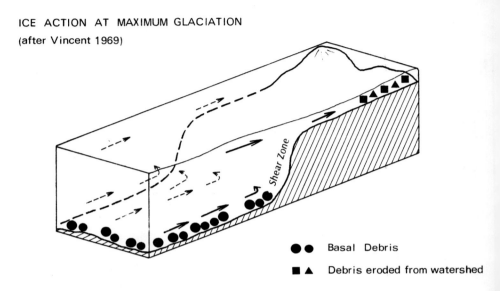

●● Basal Debris

■▲ Debris eroded from watershed

DEBRIS DISTRIBUTION AFTER DEGLACIATION

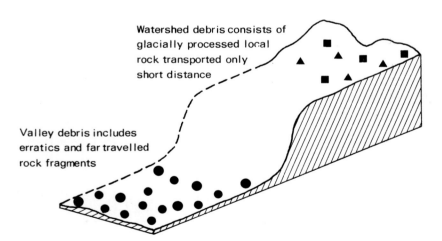

Watershed debris consists of glacially processed local rock transported only short distance

Valley debris includes erratics and far travelled rock fragments

Fig. 10 Vincent's model of glacier flow in upland area.

Vincent's hypothesis illustrates the relative motion of upper and lower layers of an ice sheet in an area of strong relief. In this case, the ice sheet advancing from the north into the north-facing valleys of the East and West Allen Rivers would be presented with the considerable obstacle of the Tyne/Wear watershed. Vincent's model indicates the lower zone of the ice sheet would shear and move parallel to the watershed whilst the upper layers of ice would continue their south-easterly motion. Thus the basal debris carried in the ice sheet would not cross the watershed. The upper layers of ice would, however, attack the watershed areas creating new basal ice debris on the crests and south facing slopes of the watershed. As the ice stagnated and then wasted away, the basal debris in the valleys could be the typical glacial till. On the watersheds however the debris would be slightly processed local bedrock eroded and transported a minimum distance by the ice sheet before being redeposited. It is doubtful is such material would differ drastically from the local bedrock decomposed by other means. It seems probable that in some cases the material would exhibit more characteristics of glacial till than in others. Thus there is a question of what is the point at which any single deposit becomes sufficiently altered to be reclassified as some other deposit type. To what extent must a process act in order to impose certain diagnostic characteristics on some given material. It appears from the data analysed in this study that the majority of surficial sediments in Upper Weardale exhibit the influences of several processes and very few deposits can be regarded as characteristic of only one process of deposition.

Accepting Vincent's model of the effect of strong relief on glacier flow it appears that the continuum between glacial till and decomposing bedrock is a reasonable result for Upper Weardale. The combination of glacial till and solifluction processes is also a reasonable result as any glacial deposit existing in Upper Weardale after the retreat of the ice sheet, must have been subject to modification by periglacial and hill-slope processes.

Examination of the stratigraphy at the sample sites shows that, every possible combination of factor influences is exhibited. Fig. 9 which shows a sample of these results indicates the overall complexity of the stratigraphy. The conclusion drawn from this is that there was a period of periglacial activity which resulted in considerable disturbance of the glacial tills in Weardale followed by the establishment of sub-areal weathering. This latter process continues to the present day and the downhill creep of material results in confused stratigraphy with disintegrating bedrock migrating downslope to overlie layers of till. The till itself is in

181

certain places found to overlie gelifluction and rainwashed hillslope material, probably a result of a lens of till migrating downslope to over-ride an existing accumulation of slope deposits. It is possible that the slope deposits were sub-glacial and received their characteristics as a result of the action of basalt meltwater before the till lens was deposited on top of them from within the disintegrating ice mass.

The widespread occurrence of factor 3 influence on the watersheds with the commonly occurring secondary influence of factor 1 suggests that this watershed area may indeed have been over-ridden by ice at the time of maximum glaciation. Thus the evidence from the present study indicates that earlier workers, by classifying deposits in terms of a single process have been led to an erroneous conclusion about the extent of ice-cover. The general conclusions about the extent of the various deposits are presented in Figure 4. In conclusion Figure 11 illustrates the general areas of dominant factor in the surface layers of the deposits in Upper Weardale. It should be noted that earlier maps of glacial till etc. are compatible with Fig. 11, however factor 1 dominance indicates either a glacial till or an altered glacial till because mapping the dominant factor does not permit any indication of secondary influences. The extension of the factor 1 influence to the watershed in the north-western part of Upper Weardale is in general agreement with Vincent's (1969) conclusion that ice may have moved southwards into the valley of the River East Allen and crossed the watershed into Weardale. The dominance of factor 1 in the surface layer here indicates that this may have been an important local ice movement resulting in the creation of a recognisable basal till. The secondary influence of factor 1 in the adjacent areas of factor 3 dominance indicates that ice may have been present on these adjacent watershed areas.

It is thus apparent that the assumptions implicit in the visual processes of mapping surficial deposits have not previously been analysed. The model on which these assumptions appear to be based is similar in its structure to the factor analysis model. In Upper Weardale the particle-size distribution of the various sediments is one of the few criteria by which deposits may be objectively classified. Sedimentary structures require subjective evaluation of the relative importance of the processes acting. Provenance and mineral content for the deposits in Upper Weardale are not diagnostic properties in differentiating the deposit-types.

Thus the sedimentologists contention that the environment of deposition may be determined from particle-size data (Klovan 1966 and Solohub and Klovan 1970) seems to provide a sound basis for

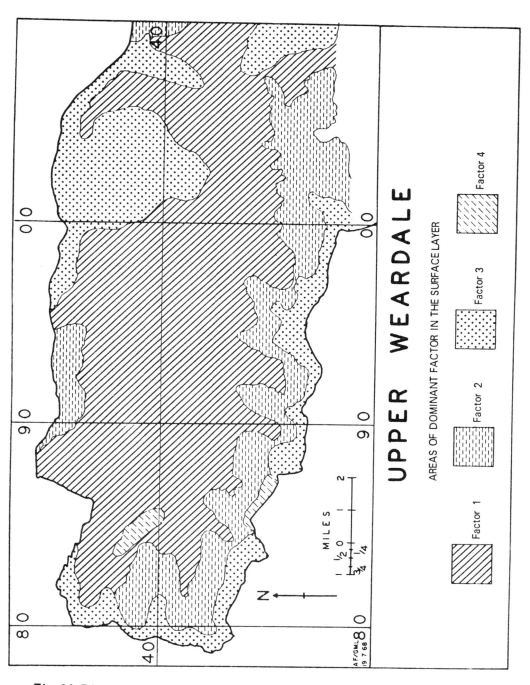

Fig. 11 Distribution of deposit types in Upper Weardale.

evaluating the suite of surficial sediments in Upper Weardale. The relative importance of the various factor influences emerges from the factor analysis of the grain-size data and the interpretation of the genetic significance of these factors remains as the major step. This interpretation, although subjective, does seem to provide a valid basis for constructing the genetic history of the deposits. It seems therefore that the technique is of considerable value in regions where surficial sediments are not stratigraphically distinct.

REFERENCES

ATKINSON, K., 1968, "An Investigation of the pedology of Upper Weardale, Co. Durham". Univ. Durham Ph.D.

BEAUMONT, P., 1967, "The glacial deposits of eastern Durham". Univ. Durham Ph.D.

DWERRYHOUSE, A.R., 1902, "Glaciation of Teesdale, Weardale and the Tyne Valley and their tributary valleys". Quart. Journ. Geol. Socl. 58 pp. 572-608

EASTWOOD, T., 1953, "Northern England". BRITISH REGIONAL GEOLOGY, H.M.S.O. London.

FALCONER, A., 1970, "A Study of the Superficial Deposits in Upper Weardale". Univ. Durham Ph.D.

IMBRIE, J., 1963, "Factor and Vector Analysis Programs for Analyzing Geologic Data". Office Naval Research Tech. Rept. 6 Geography Branch 83 pp.

JOHNSON, G.A.L., 1963, "The Geology of Moor House". H.M.S.O. London.

KLOVAN, J.E., 1966, "The use of Factor Analysis in determining Depositional Environments from Grain-Size Distributions". Journ. Sed. Petrol. 36 pp. 115-125.

KRUMBEIN, W.C. and GRAYBILL, F.A., 1965, "An introduction to statistical models in geology". McGraw-Hill, New York.

KRUMBEIN, W.C. and PETTIJOHN, F.J., 1938, "Manual of Sedimentary Petrology". D. Appleton-Century Co., New York.

MALING, D.H., 1955, "The Geomorphology of the Wear Valley". University of Durham Ph.D. Thesis.

SOLOHUB, J.E. and KLOVAN, J.E., "Evaluation of Grain-Size Parameters in Lacustrine Environments". Journ. Sed. Petrol. 40 pp. 8l-101.

TWENHOFEL, W.H., 1032, "Treatise on Sedimentation". Williams and Wilkins, Baltimore.

VINCENT, P.J., 1969, "The Glacial History and Deposits of a Selected Part of the Alston Block". Univ. Durham Ph.D.

WASHBURN, A.L., 1969, "Instrumental Observations of Mass-wasting in the Mesters Vig. District, North-east Greenland". Reitzels, Copenhagen.

MULTIVARIATE ANALYSIS OF QUATERNARY DEPOSITS USING NOMINAL SCALE DATA: ORDINATION AND INFORMATION AND GRAPH THEORETIC METHODS

J.T. Andrews

Introduction: Statement of Concepts and Problems

The advent of the high-speed computer and the development of sophisticated and powerful methods of statistical analysis has resulted, quite literally, in new dimensions being opened up to the Quaternary geologist and geomorphologist. However, it behoves us to closely examine some of the problems that might arise from the capacity to handle incredibly large matrices. In other words, I believe that we have reached a point where the capacity of the machine plus the attendant mathematics has exceeded our mental horizons. The question must be asked: If we can reduce our 100 variables to *only* 10 factors, are we in fact in a position to envisage the resulting 10 space configuration of points? The answer would have to be a definite "no". Thus one reads papers using multivariate analysis using R-mode and Q-mode factor analysis which, for visual impact and clarification, plot the loadings of the samples on Factor I and Factor II, II on III and so on. At the risk of sounding like a disciple of Marshall McLuhan, I would suggest that the majority of Quaternary scientists are conceptually restricted to a three-dimensional world and that the progress into the n-dimensional world of mathematics might result in the loss of touch between the original data and the products spilling out of the "black box". In saying this I am, of course, revealing my own inherent inability to comprehend the real physical meaning behind a method that gives me six significant factors, or even four. Possibly others are not so blind?

A second problem that I want to mention is the relationship between the original measurement and the proper statistics that can be used to evaluate the data. I subscribe to, even if I do not always strictly follow, the Stevens' dictum (Stevens 1968) that there has to be a precise relationship between scales of measurement and permissible statistical operations. The powerful and popular multivariate techniques, such as factor analysis, are quite rigorous in the measurement scales that they can use. They are restricted to measurements that are on the ratio or interval scale. They are *not* appropriate for data that is ordinal or nominal in character. These restrictions limit the uses of factor analysis because they cannot strictly accept variables that vary in their scale properties. This is not a serious objection in certain fields of interest to studies of Quaternary deposits. If we are dealing, for example, with a number of till samples and we restrict our interest to grain-size properties then factor analysis is emminently suitable. But suppose we are interested in some depositional feature and we extended our interest to include not only the mechanical properties of the deposit but we also took into account the fact that it is a three-dimensional form and hence we classify the data in terms of shape, in general size categories (large, medium, small) and also in terms of surface texture (rough, smooth) we are then faced with a different problem. By painstaking and laborious operations we might be able to define some quantitative measurement of shape, of size and of texture but in doing this we are presupposing that these measurements are worth the effort. I think we have to ask ourselves on what scale of precision do geomorphological processes operate? In terms of the stresses at the base of a glacier and in terms of the inherent topographic variability, is it likely to be critical than one drumlin is 150m and another 138m long? If size is an important variable of a deposit is it not the *relative* scale of size that is the critical factor within a study area?

Because of the operational problems attached to some of the observable properties of glacial deposits, certain potentially important variables are generally not included. Length or breadth is commonly measured, area of the deposit is less commonly included, and volume rarely at all. Similarly, shapes are classified in plan view, or cross-sectional view, but what about the true three-dimensional shape?

Properties of Glacial Deposits

An understanding of the mechanics of glacial deposition has not arrived quickly despite 100 years or so of study by numerous individuals. In part this is explained by our lack of knowledge of what happens beneath a glacier, by the absence until the last 20 years of reasonable physical models of glacier physics, and by the absence of the computer and the science of statistics of help evaluate the mass of data, and by a general unwillingness on the part of glacial geologists and geomorphologists to go beyond description into the realm of prediction. It has also been the result of a somewhat compartmentalized approach to the problem. One researcher will study the mechanical grain-size properties of a drumlin; another will evaluate the macro- and micro-fabrics; yet another might study the shape of the feature and its variations in space. Thus our knowledge is composed of a number of unrelated (unrelated in terms of cause and effect) studies on features that should be studied in their entirety--that is features with shapes, sizes, internal properties, distributions and in some relationship to the former glaciological environment. Some of the things to be measured in this "total" approach are listed on Table 1. Those used in this study on Table 2.

If we are to adopt this total approach we have to be aware of the economies of time and information. We need operational guidelines that will enable us to produce fairly quickly a body of information on a number of features. These samples should then be processed by methods that utilize nominal or ordinal scale data. The aim of this phase of the study being to evaluate the various variables, to make judgments on what is important, and then to define new, rigorous operational procedures that will aim in focusing *down* onto the genesis of the feature. This should, in other words, result in the formulation of hypotheses that are then tested (Table 3).

Analysis of Data from Cross-Valley Moraines

In the remainder of this paper I would like to discuss some different multivariate methods that can be used for this initial evaluation that do not violate any underlying mathematical restrictions. As an example, though not a complete one in terms of Table 1, of these methods, a 41 object times 12 variable array will be used. The objects are 41 sites on cross-valley moraines in north-central Baffin Island (Figure 1) and the twelve variables (or characters) are described in Table 2. The variables refer to aspects of shape, location, fabric and stone shape and roundness. Each character is made up of a number of character states, or classes, regardless of the original scale of measurement.

Cross-valley moraines occupy significant areas of Baffin Island. They are restricted to situations where the late-glacial ice retreated across a regional watershed resulting in the formation of an ice dammed lake. The moraines are restricted to the sections of the valleys that lie *below* the former glacial lake shorelines. Their formation is thus in some way related to a sublacustrine glacial movement. In the study by Smithson (1965) on these moraines the following descriptive statistics were computed (Table 4). Unfortunately, these measurements could not be related to the 41 sites used in this study. Additional descriptive information and quantitative analyses are given in Andrews (1963a and 1963b); Andrews and Smithson (1966) and Andrews and Estabrook (1971).

Ordination or Gradient Analysis

Ordination is a method of ordering samples in a three-dimensional space continuum. It was developed by plant ecologists (Bray and

. 1 Low level air photograph of cross-valley moraines of the Isortoq Valley, north central Baffin Island, N.W.T., Canada. Note the fluted surface of an asymmetric moraine in the middle foreground and the Central Kame nearby. In the foreground curved and asymmetric moraines occur side by side (photograph 1963, J. D. Ives).

TABLE 1
PROPERTIES (TO BE MEASURED) OF A GLACIAL DEPOSIT THAT ARE POTENTIALLY INDICATIVE OF ITS MODE OF FORMATION

Mechanical Properties	Fabric	Size	Shape	Distribution
Grain-size	Boulder size	length	Dimensionless ratios	clustered
Attenberg Limits	Pebble size	area	Qualitative designation	random
Density				
Clay minerology	Sand size	volume		
% constituents				
Colour				
Pebble shape				
Pebble roundness				
Lithology				

Paleoglacial Environment

Ice thickness

Velocity

Cold or wet based glacier

Age of formation

Nature of depositional environment (englacial, sublacustrine, etc.)

TABLE 2
CHARACTERS AND CHARACTER STATES OF THE CROSS—VALLEY MORAINES

Character 1:
  Moraine shape (Andrews and Smithson, 1966)
    1 = S shapes
    2 = linear
    3 = curved
    4 = asymmetric
    5 = hooked

Character 2:
  Valley position (Andrews and Smithson, 1966)
    1 = river-level
    2 = mid-valley
    3 = high-level

Character 3:
  Dip Type (Andrews and Smithson, 1966)
    1 = 1
    2 = 2
    3 = 3
    4 = 4
    5 = 5
    6 = 3 & 4
    7 = 1 & 3

Character 4:
  Orientation strength (Chi-Square test) Proximal slope
    1 = less than 35
    2 = 36 - 70
    3 = 71 - 105
    4 = 106 - 140

Character 5:
  Orientation strength (Chi-Square test) Distal slope
    1 = less than 35
    2 = 36 - 70
    3 = 71 - 105

TABLE 2 (cont.)

Character 6:
Dip strength (Chi-Square test) Proximal slope

1 = less than 10

2 = 11 - 20

3 = 21 - 40

4 = greater than 40

Character 7:

Dip strength (Chi-Square test) Distal slope

1 = less than 10

2 = 11 - 20

3 = 21 - 40

4 = greater than 40

Character 8:

Ratio: $\dfrac{\text{No. in proximal mode dipping up glacier}}{\text{No. in distal mode dipping up glacier}}$

1 = less than 1

2 = 1.1 - 2.0

3 = 2.1 - 4.0

4 = greater than 4.1

Character 9:
No. of ovoid stones Proximal slope

1 = less than 4%

2 = 4.1 - 8.0%

3 = 8.1 - 16.0%

4 = greater than 16.1%

Character 10:
No. of ovoid stones Distal slope

1 = less than 4%

2 = 4.1 - 8.0%

3 = 8.1 - 16.0%

4 = greater than 16.1%

TABLE 2 (cont.)

Character 11:

$$\frac{\% \text{ angular stones}}{\% \text{ subrounded stones}}$$ Proximal slope

1 = less than 1:2.0

2 = 1:2.1 - 1:5.0

3 = 1:5.0 - 1:15.0

4 = 1:15.1 - 1:30.0

5 = greater than 1:30.1

Character 12:

$$\frac{\% \text{ angular stones}}{\% \text{ subrounded stones}}$$ Distal slope

1 = less than 1:2.0

2 = 1:2.1 - 1:5.0

3 = 1:5.1 - 1:15.0

4 = 1:15.1 - 1:30.0

5 = greater than 1:30.1

TABLE 3

DESIGN OF A STUDY OF GLACIAL DEPOSITS

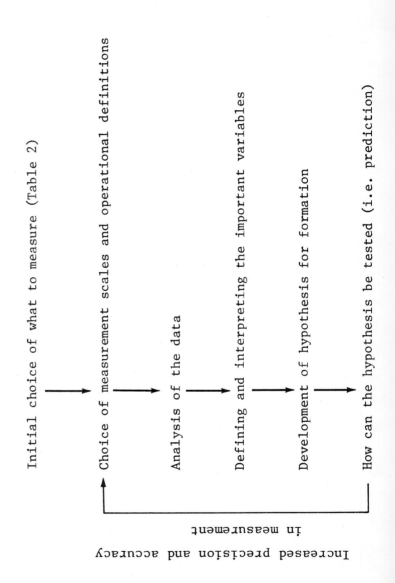

TABLE 4

SELECTED CHARACTERISTICS OF THE CROSS-VALLEY MORAINES

N = No. in sample

	Proximal slope			Distal slope		
	X	S	N	X	S	N
Slope angle						
(upper section)	18.3°	5.6	114	23.5°	6.0	122 +
(lower section)	16.1°	6.0	111	22.4°	6.0	119 +
*Height (m)						
(upper level)	2.53	1.65	110	2.56	1.5	119
(lower level)	2.69	1.73	113	2.60	1.5	122
*Spacing						
(upper)				32.6	25.2	120
(lower)				33.9	27.6	117
Penetrometer ($kg\ cm^{-2}$)	13.07	3.67	26	11.2	3.6	25 +
Grain-size (phi units)	1.53	0.43	19	1.30	0.35	18 +

* Skewed distributions

+ Different at 95% significance level, "t" test.

Curtis, 1957) who were seeking methods to express the concept that plants are related to each other in a space continuum rather than representing a series of discrete clusters (i.e. plant communities). The method plots samples in terms of X, Y and Z cartesian co-ordinates that are derived from a similarity and dissimilarity matrix (see below). The approach has been used in other ecological studies but it has had limited exploitation in the physical sciences. An ordination of soils was developed by Hole and Koronaka (1960) and Feldhausen (1970) has more recently used it to describe the sediments off Cape Hatteras.

The material used in this paper was obtained from a computer program written by Reid and Webber and the Institute of Arctic and Alpine Research, University of Colorado (Reid and Webber, *in press*). Output is in the form of a microfilm plot that can be used as a slide or it can be used to print final figures.

Procedure:
Sorenson's coefficient was calculated for each paired comparison. The coefficient is also called "Percent Similarity" (PS) and is defined:

$$PS = \frac{2w}{(a + b)} \times 100 \qquad \ldots (1)$$

where w is the sum of the lowest value for each paired comparison, and a and b are the sums of the states of each variable (Table 5). A matrix of the PS coefficients is set up and at the same time the disimilarity coefficient is obtained from DS = 100 - PS. For each site or sample the PS and DS values are summed and these are used to select the reference samples. In the case of the cross-valley moraines, moraines   5 and   40 are most different with a disimilarity coefficient of 42.978. The X axis has as end points   5 and   40 and they are scaled along a line 42.978 units in length. All the other 39 samples are then placed on this line by a simple distance algorithmn. Sites half-way between #5 and #40 are most unlike the X axis reference samples and they are then used to define a Y axis, in this case with end points #15 and #14 and separated by 32.833 units. Samples most unlike the X and Y, reference samples occur in a central location in the middle of the X-Y plane, these are then used to define the final Z axis (samples #1 and #9 and #29 which are 25 units apart). The final outcome is thus the calculation of X, Y and Z co-ordinates for the 41 samples; the samples are now ordered within a cube based on figures given in Table 6.

In the program developed by Reid and Webber, the ordination is drawn on a microfilm plotter so that the results can be used

TABLE 5

CALCULATION OF PS

Variable

Object	1	2	3	4	5
1	2	2	4	3	2
2	1	2	1	2	1

$$w = 1 + 2 + 1 + 2 + 1 = 6$$

$$a = 2 + 2 + 4 + 3 + 2 = 13$$

$$b = 1 + 2 + 2 + 2 + 1 = 7$$

$$PS = \frac{2(6)}{13 + 7} \times 100 = 60\%$$

TABLE 6

X, Y, Z CO-ORDINATES FOR THE MORAINE ORDINATION

Site Number	X-Axis	Y-Axis	Z-Axis
5	0.000	9.328	12.959
7	12.524	14.648	14.564
25	14.835	22.268	5.881
2	17.939	3.581	7.571
6	20.278	24.756	11.465
24	20.620	25.892	7.070
20	22.705	6.080	9.342
26	22.725	17.473	18.760
21	22.884	14.499	18.157
3	23.192	17.446	7.590
11	23.246	14.982	0.000
30	23.275	14.134	17.876
1	23.482	17.536	3.612
23	23.762	32.833	10.621
8	25.018	0.000	11.324
28	26.405	14.445	22.552
29	26.405	14.445	25.081
9	27.047	9.838	25.081
4	28.426	12.702	6.004
16	28.724	9.141	14.509
33	29.175	14.283	23.041
32	29.773	10.760	11.855

TABLE 6 (cont.)

Site Number	X-Axis	Y-Axis	Z-Axis
27	31.225	2.396	21.857
10	31.486	14.871	-3.793
18	31.643	4.345	8.749
14	32.426	1.021	7.571
38	32.769	10.319	11.214
19	33.398	17.833	18.601
22	33.412	8.489	14.713
31	33.942	.084	14.645
17	34.870	.431	6.994
15	35.645	-.584	2.947
13	36.055	3.581	9.170
39	36.055	7.684	11.544
34	36.102	5.582	22.585
35	37.312	-.767	14.532
36	37.810	.719	9.802
12	38.020	3.195	10.621
37	40.183	7.684	14.159
41	41.373	8.067	14.377
40	42.978	7.684	18.373
Distance Between Reference Stands	42.978	32.833	25.081
Reference Stands	5-40	8-23	11-29

directly as 35mm slides or good quality photographic enlargements can be (as in Figures 2, 3 and 4) produced. The ordination is plotted as three separate diagrams, namely the points on the X-Y, and X-Z and Y-Z planes. Webber (person, commun. 1970) has also constructed a model of the ordination using a three-sided box with individual points inserted in the form of thin wires. This enables the researcher to completely visualize the spacial quality of his data.

A germane question at this stage in the analysis is the relationship between the ordination and the original variables. The computer program prints out the location and magnitude of each variable on the three planes (Figures 2, 3 and 4). If there is some relationship between the ordination and the variable, the sites of a particular class should occupy the same area of the cube and be distinct from other classes. Figure 4 is a good example of this. The character is #11 (Table 2) which is the percent of angular stones/percent of subrounded stones. There are five classes of this measure and they are shown by squares whose size is proportional to the class name. There is some overlap between these classes when plotted against the ordination but nevertheless the classes are moderately distinct. Such relationships indicate that the ordination would correlate significantly with such a variable and such variables might be considered as controlling variables. In terms of this study, good separation is achieved by variables 1, 2, 9 and 11 (see Table 2 for explanation). Till fabric characters are not well related to the ordination although some detail is apparent (Figure 5 A and B). It is possible to include most of classes 2 and 3 within boundaries that do not overlap. They are also placed such that there is a graduation from a middle area of class 1 through classes 2 and 3 with the fabrics with very high Chi-square values occupying the outer portions of the ordination. A poor relationship would be one in which the classes were completely mixed within the 3-space.

Of particular concern in attempts to understand the genesis of cross-valley moraines is the relationship between the ordination and moraine shape. How clearly do the various attributes of fabric strength, stone shape and roundness and moraine position agree with the morphological expression of the moraine? Does morphology reflect structure is perhaps the condensed form of this query? Only three moraine types were used in this study whereas a previous publication used five (see also Table 2). Type 1 is composed of simple linear and S-shaped moraines; Type 2 is the asymmetric moraines of Andrews and Smithson (1966) and Type 3 is hooked and curved moraines. Figure 6 represents the ordination on the

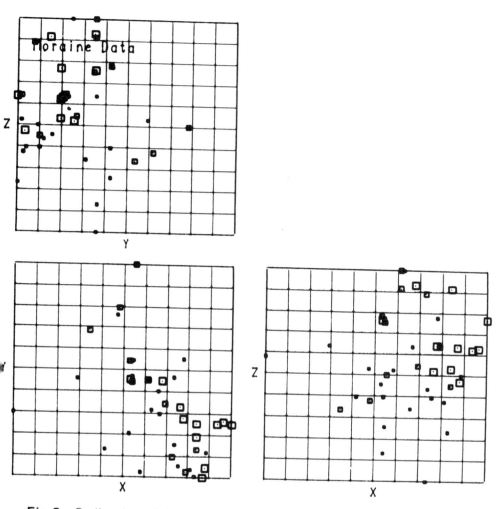

Fig. 2 Ordination of the moraines against three moraine types (See also Figure 6).

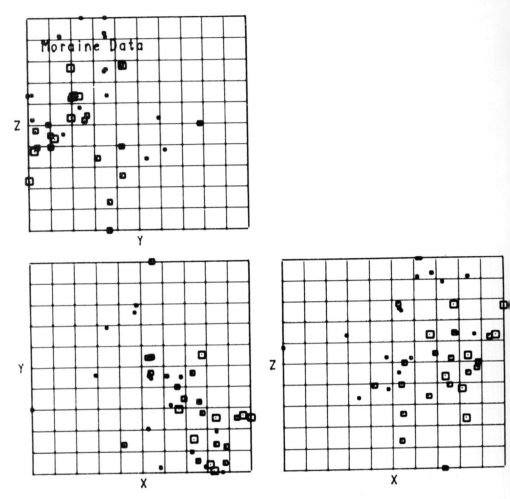

Fig. 3 Ordination against valley location.

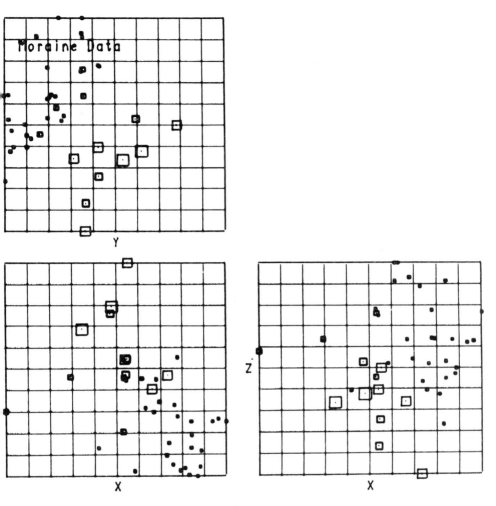

Fig. 4 Ordination against Character 11 (Table 2).

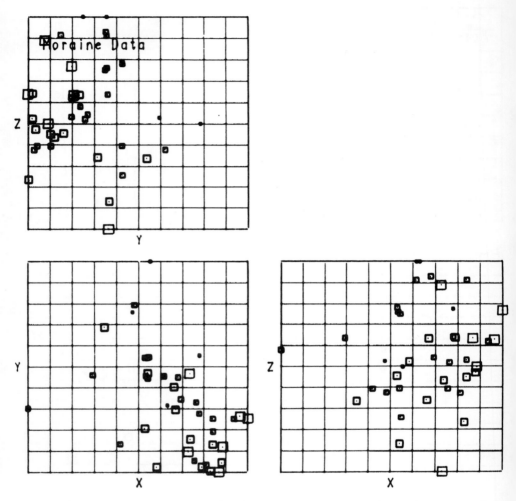

Fig. 5a Ordination against till fabric strength on the proximal slope.

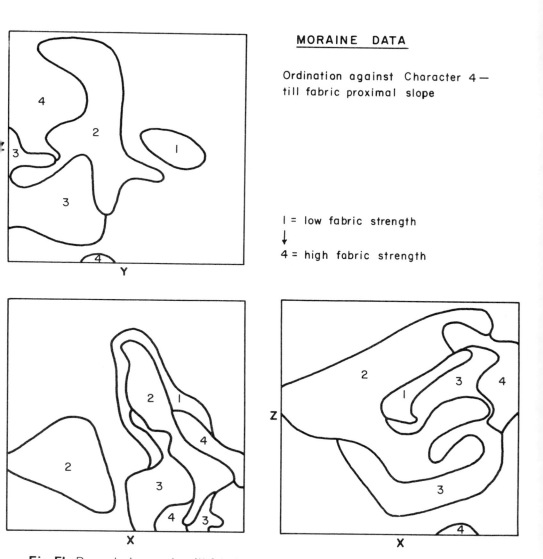

MORAINE DATA

Ordination against Character 4 — till fabric proximal slope

1 = low fabric strength
↓
4 = high fabric strength

Fig. 5b Boundaries on the till fabric classes from Fig. 5A above (see Table 2 for class limits).

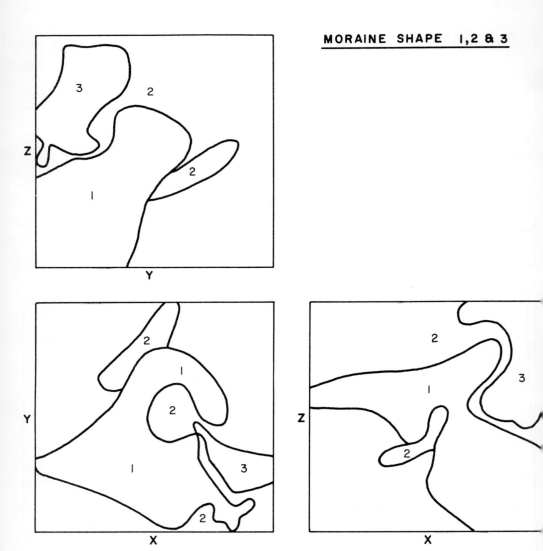

Fig. 6 Boundaries on moraine type as delimited on the ordination of Fig. 2.

three types. The hooked moraines (#3) form a very compact cluster in comparison to the simple linear moraines which, although distinct as a group, cover considerably more "space". Sandwiched between moraine types 1 and 3 are the asymmetric moraines. These lie not only between types 1 and 3 but three lie in an isolated group, quite distinct from any type 3 moraines. As the moraine types were one of the variables they do have some weight in the final ordination, but only 1/12th in this example so it is unlikely that their inclusion has overly influenced the final product. The analysis indicates that the simple linear moraines and the hooked moraines are, on the average, more different than either with the asymmetric moraines. This result was not expected.

The reasons are not fully understood but may relate to the preferred location of the different moraine types in the valley (Table 2, character 2) and the complex interaction between till fabrics, matrix, clast shape and roundness. From a purely mechanical point of view, position must be important as this controls the magnitude of the stresses that led to the formation of the moraines. A location near to the glacial lake shoreline implies:

1) low shear stresses at the base of the ice because the ice is thinner in this location,
2) limited depth of water,
3) coupled with 1) above and possibly also a function of englacial temperatures is the probability that ice velocity was less than in the center of the valley.

The ordination against valley position is shown as Figure 3. Some of these more detailed questions can be answered through information- and graph- theoretic methods outlined below.

Information- and graph-theoretic methods

Information- and graph-theoretic methods of analysing nominal and ordinal scale data were developed for use in Numerical Taxonomy by the former Taximetrics Laboratory at the University of Colorado. The methods are being increasingly employed by Quaternary geologists and geomorphologists at that University with very encouraging results. A discussion of the application to Quaternary deposits (the cross-valley moraines) have been recently published (Andrews and Estabrook, 1971). Other completed or on-going applications include: factors effecting corrie glacierization (Andrews and Dugdale, 1971); a similar study on the distribution of rock glaciers in the San Juan Mountains of Colorado (Andrews and Carrarra) and two studies on relative moraine chronologies using various weathering criteria (Miller and Yount).

i) Information theory approach (analagous to R-mode approach):

Two characters, with different character states or classes, can be said to share information if knowledge about one enables probability statements to be made about the other (Estabrook, 1967). If knowledge of one character tells us nothing about the other then the characters share no information. In general, information in two characters can be thought of as a Venn diagram with the degree of overlap being the amount of shared information. The concept is explained diagrammatrically on Figure 7. A program called CHARANAL computes the information in a character, the amount of shared information, a measure of distance, D, such that identical characters have D = 0, and conditional and unconditional probability tables.

The value for D for any character pair X and Y is defined (Figure 7) by:

$$D(x,y) = \frac{H(x/y) + H(x/y)}{H(x/y) + H(x/y) + R(x/y)} \qquad \ldots (2)$$

The amount of information in a character is governed by the number of objects in the study and by their distribution through the character states. For example, the amount of information (H) in a character with three character states and proportion of objects such that character states 1 = 0.4146, 2 = 0.3902 and 3 = 0.1951 is H = 0.4146Ln (0.4146) + 0.3902LN (0.3902) + 0.1951Ln (0.1951) = 1.5164. Here Ln stands for the log to the base 2.

From the output of CHARANAL, matrix tables of Distance (a symmetric matrix) or shared information (unsymmetric) can be developed (*see* Andrews and Estabrook, 1971). They can be used to "sort" the characters into three groups--those that contribute little or nothing to the study (low percent information and high D values); those that are redundant (because they are too similar to other characters); and the remainder that are those that might contribute something to a classification. In the study of the cross-valley moraines, the characters are ranked from highest to lowest in terms of shared information as: 9, 11, 12, 3, 10, 8, 1, 5 and 6, 2 and 4. In this particular study, all the variables had higher average shared information than a group of "dummy variables" that were included which had been derived from random number tables.

For detailed interpretation the conditional probability tables that are outputed from CHARANAL provide considerable insight into the stochastic nature of the data. This is considered in the paper by Andrews and Estabrook (1971) and will not be discussed here.

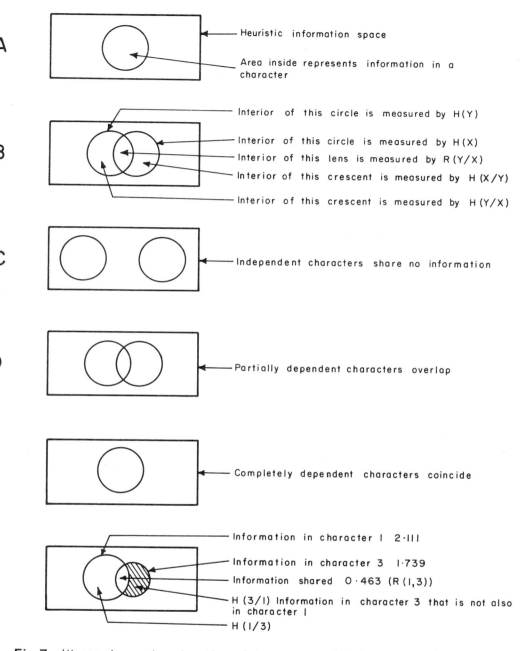

Fig. 7 Illustration and explanation of the concept of "information space", distance and fraction of shared information.

At this point let us examine possible ways of analysing nominal scale data in terms of extracting clusters of variables that are similar. This might be considered analagous to extracting the "factors" in a conventional R-mode routine. It can be done in two ways--the first would be by simply rotating the data matrix through 90° so that the variables are now considered as the objects. In this particular study the matrix now consists of a 12 object times 41 variable array. An ordination can now be performed to see how the characters of moraine shape, position etc. (Table 2) are positioned in our three-dimensional reference space. The same approach could also be used using the GRAPH program (Andrews and Estabrook, 1971). This program computes a similarity measure, c, and then proceeds to order, or partition, the data. Part of the output is a visual impression of the relationship amongst the data in terms of a "skyline" graph (Figure 8). Another approach is used in this paper--the shared information matrix was used to rank and score the degree of association between one variable and all others. The ranking was such that shared information of $\leq 10\% = 1$, 11 to 20% = 2 and so forth. Figure 8 shows the grouping of the moraine variables: Two distinct clusters are obvious at the 0.5 similarity level with the first cluster composed of two sub-clusters. Moraine shape, dip pattern and characters #9, and #11 are grouped together whereas the other cluster links #4, #6, #7, #8, #10 and #5. The link variable between the two clusters is #2, that is the position within the valley. The importance of this variable was also stressed earlier (Andrews and Estabrook, 1971).

ii) Graph theory methods:

Once the characters have been analysed in terms of their information content and their significance to the study, the next step is to examine the groupings of "similar" moraines as defined by the characters being considered. At this point we should remember that so-called "objective classifications" are a figment of the imagination. The computer might determine the actual groupings or clusterings but the rationale that lies behind the choice, and significance of the variables, is the professional judgment of the investigator and it is this which really controls and manipulates the final outcome.

Most clustering methods have at least two problems to my mind, the first is that there are few measures of the compactness of a cluster; and the second problem is that all the variables are given the same weight even though field experience might suggest that the variables differed in importance one from another.

210

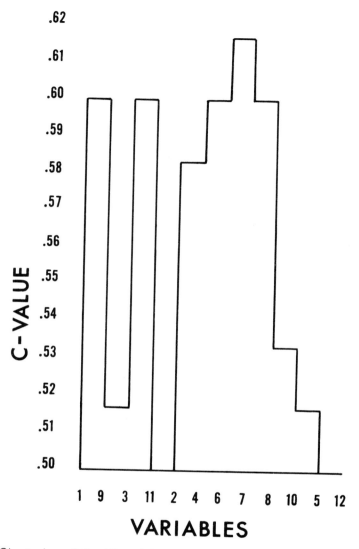

Fig. 8 Clustering of the 12 variables based on the amount of shared information using the "skyline" plot from GRAPH.

211

A measure of compactness is provided GRAPH. If at some level a cluster is delimited there is a theorectical number of links between members of the cluster. This is given by $n(n-1)/2$ where n is the number of objects in the cluster. This figure is computed and is compared with the actual number of links. The ratio of: actual/theoretical then gives an adequate measure of the internal similarity of that cluster. In general clustering methods provide only a superficial impression of the interrelationships amongst objects as data on the degree of internal connection between cluster members is not often available on most conventional programs.

The weighting of the variables to express a degree of professional judgment and opinion has potential promise although it is not without its problems. It does presuppose that the investigator has considerable insight into the processes that he is concerned with. However, it can be argued that an equal weighting of all variables is unreasonable. One suggestion is that the variables can be ranked according to their average amounts of shared information and that weightings be established on this basis. If the variables are ranked from 1 to 9 (highest to lowest) the character which has most information might be repeated 9 times, the next 8 etc. Thus the number of variables would total 45. GRAPH could now be run on this data matrix. At the moment we are still experimenting with this approach.

One final comment concerns the actual calculation of the similarity measure by the GRAPH program. Most distribution-free similarity measures are based on a simple yes/no criterion. On-the-other-hand, the similarity measure used here (Estabrook and Rogers, 1966) allows for three different types of character state relationships. The first is the usual yes/no type so that character state 1 is considered different from character state 2. However, if the character states are ordered (e.g. small, medium, large) the researcher might like to express the notion that there is a degree of similarity between character states 1 and 2, and 2 and 3 but not between 1 and 3. This can be done in GRAPH. Finally, the researcher might be dealing with a complex ordering so that state 1 is more like state 4 then 1 is to 3. A good example of this type of data is orientation data divided into 45° sectors. Such a character is termed a matrix character and it can be accommodated in the calculation of the similarity measure used in GRAPH.

SUMMARY

Methods have been described that handle data at the nominal/ordinal level of measurement. They have considerable applicability to studies on Quaternary deposits, particularly at the level of a "first look" where a large qualitative data matrix is analysed for those variables that are judged important by the investigator. I have run experiments to judge the loss of basic information between factor analysis of ratio data and Information-theoretic analysis of the same data grouped into classes. The results indicate that the same relative ranking of the variables is maintained.

The absence of any formal statistical tests on the results is at first sight a drawback, but this can be alleviated by the development of dummy variables that are then run through CHARANAL and GRAPH. The resulting values of D and c form normal distributions and significance levels can then be attached to the actual analysis. For eight character states per character c values of > 0.5 are probably non random at the 99% significance level.

We are using ordination and the information and graph-theoretic methods at the University of Colorado on a number of Quaternary studies. Examples include an ordination of corrie glaciers from East Baffin Island, N.W.T. that was very successful; examination of the factors controlling the presence of rock glaciers in the San Juan Mountains of southwest Colorado, and the relationship of different clay minerals to age in the eastern mountains of Baffin Island. In this last study the clay mineral quantities are ranked as: absent, trace, small, moderate, large.

ACKNOWLEDGEMENTS

P.J. Webber and W. Reid of the Institute of Arctic and Alpine Research and Department of Biology, University of Colorado provided valuable discussion on the problems and use of ordination. Their program was used in this study. CHARANAL and GRAPH were developed by the Taximetrics Lab., University of Colorado. Funds to run the programs was provided by the Computing Center at the University. A much faster version is now available than the one quoted by Estabrook and Andrews, 1971.

REFERENCES

Andrews, J.T., 1963a: Cross-valley moranines of the Rimrock and Isortoq River valleys, Baffin Island, N.W.T.: a descriptive analysis. *Geog. Bull.*, no. 19, pp 49-77.

Andrews, J.T., 1963b: The cross-valley moranines of north-central Baffin Island: a quantitative analysis. *Geog. Bull.*, No. 20, pp 82-129.

Andrews, J.T. and R.E. Dugdale, 1971: Factors affecting corrie glacierization in the Okoa Bay region, East Baffin Island, Canadian Arctic. *Geol. Sco. Am. Abstracts,* Vol. 3(4), pp 253.

Andrews, J.T. and G.F. Estabrook, 1971: Applications of information and graph theory to multivariate geomorphological analyses. *J. Geol.*

Andrews, J.T. and B.B. Smithson, 1966: Till fabrics of the cross-valley moraines of north-central Baffin Island, Northwest Territories, Canada, *Bull. Geol. Soc. Am.,* Vol. 77, pp 271-290.

Bray, J.R. and J.T. Curtis, 1957: An ordination of the upland forest communities of southern Wisconsin. *Ecol. Mon.,* Vo. 27(4), pp 325-349.

Estabrook, G.F., 1967: An information theory model for character analysis. *Taxon,* Vol. 16, pp 86-97.

Estabrook, G.F. and D.J. Rogers, 1966: A general method of taxonomic description for a computer similarity measure. Bio Science, Vol. 16, pp 789-793.

Feldhausen, P.H., 1970: Ordination of sediments from Cape Hatteras continental margin. *Math. Geol.,* Vol. 2, pp 113-129.

Hole, F.D. and M. Hironaka, 1960: An experiment in ordination of some soil profiles. *Proc. Soil Soc. America,* Vol. 24, pp 309-312.

Reid, W. and P.J. Webber, in press: A computer ordination program. *Arctic and Alpine Res.*

Smithson, B.B., 1965: The glacial geomorphology of the upper Isortoq River, Baffin Island, N.W.T. MSc thesis, Univ. Western Ontario, London, Ont. 153 pp.

Stevens, S.S., 1968: Measurement, statistics and the schemapiric view. *Science,* Vol. 161, pp 849-856.

THE USE OF TREND SURFACE ANALYSIS IN THE INTERPRETATION OF QUATERNARY DEPOSITS

H. Olav Slaymaker and Michael Church

The geomorphologist is characteristically involved in the interpretation of spatially distributed landform elements which may often be considered as the response features of an environmental model (Krumbein and Sloss, 1963). Process elements of the Quaternary environment have largely disappeared and the former existence of ice sheets, valley glaciers, pro-glacial lakes and the like must be inferred from the properties of the response elements which remain. These may be, for example, the fabric of a till or the size and sorting of sedimentary particles, where we have confidence that no major sedimentary transport processes have subsequently modified the response elements.

The characteristics of each response element vary over space and it is possible to construct maps of various properties of the response elements by simply interpolating isopleths amongst observed values. Such a map will direct attention to any gross regional pattern that may exist and to some of the more extreme anomalies. It will fail to distinguish many of the detailed characteristics of that pattern and any local anomalies. In order to gain greater resolution, each measured property of the response element may be considered as a dependent z-value in the triplet (x, y, z), where the x and y values are the co-ordinates of location. The grouping of z-values is then interpreted as a smooth response surface, and the surface with the best fit to the x, y and z co-ordinates may be generated.

A "trend surface" is a regularly varying surface that portrays the general pattern or trend of a set of data distributed in 2-space or higher dimensioned spaces (trend hypersurfaces). Amongst the methods commonly used to determine the trend (Rao and Rao, 1969) are smoothing (method of moving average, cf. Beloussov, *et al.,* 1965), relaxation (cf. Paul, 1967), and least squares fitting. The present paper will concentrate on the last of these.

The best fit surface is fitted in three-dimensions in much the same way as a least squares regression line is fitted in two dimensions (cf. Drumbein, 1959; Krumbein and Graybill, 1965). The total variance about the mean surface elevation of the dependent z-values $(z_{iobs} - z)$ is partitioned into two components: that subsumed by ("explained by") the best fit surface $(z_{icomp} - z)$ and that due to residual variations $(z_{iobs} - z_{icomp})$ about that surface.

The explained and unexplained variances are commonly understood to represent variability on at least two levels (cf. Miller and Olson, 1955; Krumbein, 1956, 1958). The explained variance measures the regional variability: the unexplained variance, corresponding to residuals from the trend of the response surface, results from variability at lower levels. Thomas (1960), Whitten (1959, 1960), and Merriam (1964) have given considerable attention to patterns of residuals as representing the effect of purely local variability in data. Whitten (1963) also discussed the discrimination of primary (regional) and secondary (local) trend components by comparing low and high order trend surfaces (cf., also, Allen and Krumbein, 1962; Cook, 1969). In problems about depositional environments it is generally held that these large and small scale variabilities are attributable to large and small scale sedimentary processes. The separation of mapped data into at least two main parts representing large and small scale responses is an important characteristic of the form of process-response models that is entailed by trend surface analysis.

The problem of determining the trend in spatially distributed data was first broached, in a practical way, by geophysicists, who sought to distinguish the regional pattern and local anomalies in gravity data (cf. Agocs, 1951; Simpson, 1954; Oldham and Sutherland, 1955). If it is desired to proceed beyond simple plane trends, or to subsume large amounts of input data, considerable computational power is required, and so the extensive development of trend surface methods only occurred after computers became available for use.

Trend surface methods that involve least squares fitting of an explicit functional form can be divided into three general classes: orthogonal polynomials (Grant, 1957; Whitten, 1970), non-orthogonal polynomials (Krumbein, 1959; Mandelbaum, 1963), and Fourier functions (Swartz, 1954; James, 1966). Agterberg (1964) and Krumbein and Graybill (1965) give detailed discussions of the polynomial methods, and Krumbein (1966) has compared all three approaches. James (1967) has considered more complex, non-linear functions. Computer programmes for application of the techniques have been extensively developed at the Kansas State Geological Survey, whose work is summarized by Harbaugh and Merriam (1968). Advanced applications of trend surface analyses include those of Krumbein and Imbrie (1963) in association with factor analysis, Harbaugh (1964) in higher-order hyperspaces, Sneath (1967) in the study of transformation grids, and Agterberg and Cabilio (1969; also Agterberg, 1970) in multivariate prediction.

The detection of significant trend components in noisy data has been considered by Allen and Krumbein (1962), Chayes and Suzuki (1963), Howarth (1967), McIntyre (1967) and Tinkler (1969). Because of the flexibility of higher-order trend components, care must be taken in interpreting the significance of fitted trends. The sensitivity of trend analysis techniques has received little direct attention (but cf. Romanova, 1970). Allen and Krumbein (1962; also Krumbein, 1963) have considered the establishment of confidence limits about low-order surfaces.

Geologists have been the leaders in adopting trend surface techniques as a descriptive and interpretive tool (cf. Krumbein, 1956; 1959; 1963; Miller, 1956; Svensson, 1956; Whitten, 1959; 1960; 1961). Students of Quaternary sediments have not been innovators in technique, but nevertheless a body of results exists which indicates that trend analysis methods are being appreciated and adopted to help solve problems of Quaternary sediments. Three major applications can be seen illustrated in the following examples.

The Mean Properties of Surfaces

It is often of interest to isolate a spatially varying mean characteristic of some surface on a regional scale. When the initial data are relatively noisy, the computation of trend surfaces provides a means for determining the mean characteristic as a regional trend, and for assessing its significance.

A study which attempts to analyse regional patterns of till fabric orientation is provided by Roberts and Mark (1970). They computed trend surfaces on mean orientation data and then used the derived regional trends to construct the indicated ice flow lines (cf. Figure 1). The technique is potentially very powerful for defining ice sheet movement in areas such as eastern England as they showed using Baden—Powell's (1948) conclusions and data from West and Donner (1956). Its usefulness in high relief areas is more open to question. However, by using nineteen sets of fabric data for tills of the Sumas stage in the Mission-Abbotsford area of the lower Fraser Valley of British Columbia, they derived a trend surface showing a regional north-south trend, with a presumed northerly ice source (Figure 2). Armstrong, Matthews and Sinclair (personal communication) have questioned the results of the analysis largely on the grounds that other evidence, notably the provenance of the till lithology, contradicts the trend surface results. They suggested partitioning the data and fitting two trend surfaces, ultimately producing an interpretation of the most likely flow pattern based on consideration of provenance, inferred ice

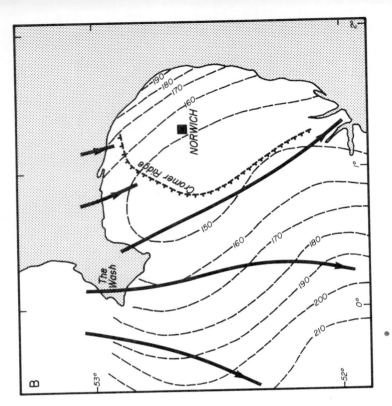

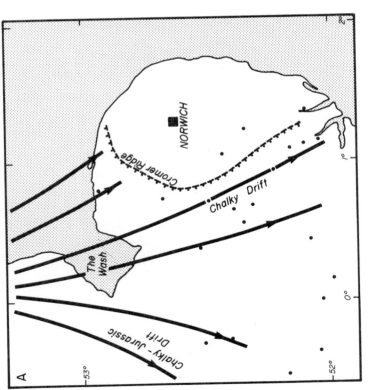

Fig. 1 Eastern England

A. Ice flow lines for Gipping Glaciation as proposed by Baden-Powell (1948) with West and Donner's (1956) till fabric sites indicated.

B. Isoazimuths of third order trend surface fitted to Gipping Till fabrics of West and Donner (1956) and ice flow lines derived by Roberts and Mark (1970).

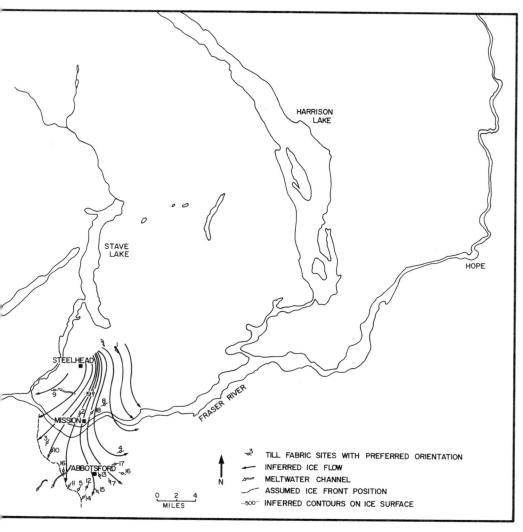

Fig. 2 Fraser Valley. Ice flow lines for Sumas Stage derived from third order trend surface (Roberts and Mark, 1970) with 19 till fabric sites indicated.

front position and till fabric data (Figure 3). It is not the purpose of the present discussion to choose between Figures 2 and 3 as the better reconstruction. It seems to us that a fundamental weakness intrinsic to both as far as the application of trend surface analysis is concerned is that in this high energy environment the regional trend is less important than local variability. As the forms of the analyses employed specifically avoid the analysis of residuals, only limited success with the application of this technique can be anticipated in high energy environments.

Identification of Significant Scales of Variability

In some studies it is important to distinguish the various scales of variability in data. In this context trend surface analysis can be used to identify the regional component and to isolate the residual variance at more local scales for further investigation.

A study which attempts to establish regional trend and also analyses variability in the residuals is provided by Chorley *et al* (1966). The Breckland of eastern England is an area of sandy soils, overlying an undulating till plain (Figure 4). The sandy soils are generally held to be decalcified till, whose constituents originated in the Cretaceous Sandringham Sands. Work by Baden-Powell (1948) and by West and Donner (1956) used till fabric data to infer directions of ice sheet movement during the Pleistocene (cf. the preceding section, also). However, no work on establishing the mean size characteristics of the sedimentary response surface nor on the local variability of that surface had previously been reported.

One hundred and forty-nine systematic sampling sites were located at two kilometer intervals. One kilogram samples from the twelve inch layer immediately below the humus layer were taken and, by mechanical analysis, median sizes were determined. A best fit cubic trend surface was fitted to the D_{50} data, for which the "explained" variance was 21.5%. This was statistically significant at the 95% level, and the form of the surface could be rationalized in terms of known ice movements in that the coarser sands where shown to be in the northeast and the finer sands in the southwest (Figure 4). Nevertheless, a high proportion of the total variance remained unexplained.

The 78.5% unexplained variance was analysed in terms of a hierarchical model of variance, involving six distinct sampling levels. By examining the distribution of residuals from the cubic surface, two areas, A and B, each four kilometers square, the one with strong negative residuals in the southwest and the other with strong

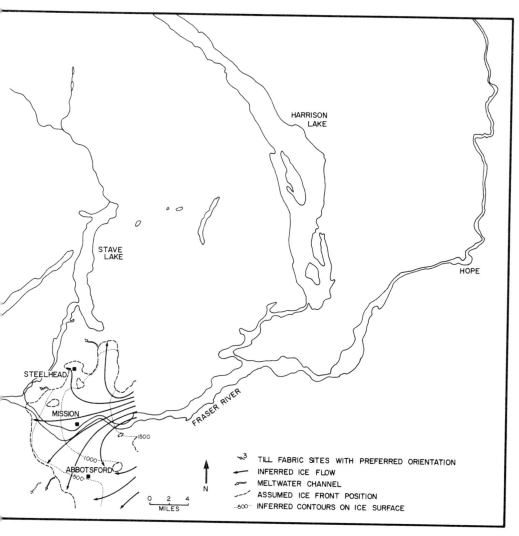

Fig. 3 Fraser Valley. Ice flow lines for Sumas Stage derived from (a) provenance of till lithology (b) inferred ice front position and (c) two third order trend surfaces (Amstrong, Mathews and Sinclair, pers. comm.).

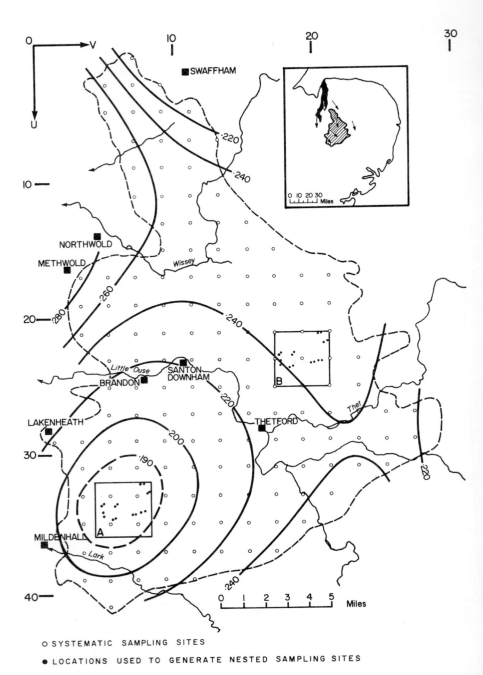

O SYSTEMATIC SAMPLING SITES

● LOCATIONS USED TO GENERATE NESTED SAMPLING SITES

Fig. 4 Location map of the Breckland of Eastern England. Third order trend
surface fitted to median grain sizes (in mm) of surface Breckland
sands at 149 systematic sampling sites. (after Chorley et al 1966)

positive residuals in the northeast (Figure 5) were defined. Sixty-four sampling sites within each of the two areas A and B were then located according to a hierarchical sampling design and D_{50} was determined for all 128 sites. The variance component estimate (Krumbein and Slack, 1956) for each of six hierarchical levels was then determined to isolate the greatest sources of variability. Figure 6, by plotting cumulative variance component estimates for three of the six levels (for all 128 samples, for the 64 samples in A and B separately, and for the 16 samples in four quadrats of A and B separately) suggests that most significant variability is introduced at the 17 kilometre, the 245-826 metre and the 8 metre levels. These correspond to a "periodic" structure in the areal pattern at a characteristic spacing. The 17 kilometre spacing is associated with regional sand facies variation described by the trend surface. The 245-826 metre spacing corresponds to possible structures superimposed on the regional trend with diameter between about one-eight of a kilometre to about one kilometre. The eight metre spacing corresponds to structures superimposed on these latter structures, probably some kind of periglacial stone polygons or stripes. (Photographs in Chorley et al, 1966).

The conclusions deriving from the partitioning of variance bore little relationship to the trend surface analysis and, contrary to the findings on Baffin Island (see below), suggested that local variability was not closely related to regional variation in grain size. This may be partly due to the age of the till (Gipping) and the large number of subaerial processes which have modified it during the Hoxne Interglacial and the Hunstanton Glaciation.

In this example, trend surface analysis was used not only to establish a regional trend but, more importantly, to locate the sampling quadrates A and B by careful analysis of the distribution of residuals from the surface, such that the most pronounced regional differences in median grain size were included in the sample.

Inferences About Physical Processes

Trend surfaces may be used to support inferences about the nature of physical processes. The response that is predicted from the operation of certain processes can be checked by inspecting the form of the computed trend surface.

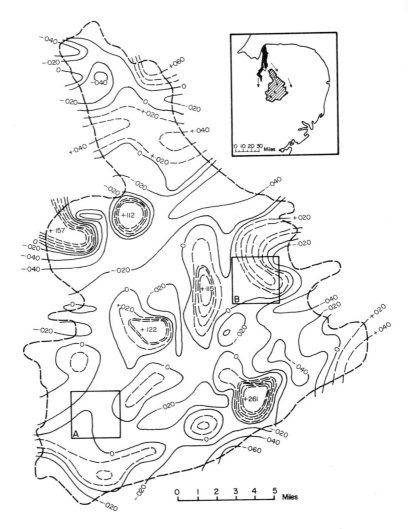

Fig. 5   Breckland. Residuals from the third order trend surface.
(after Chorley et al 1966)

224

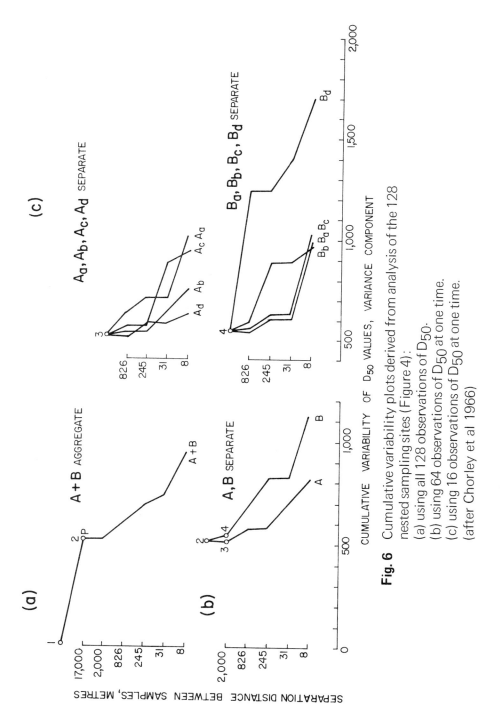

Fig. 6 Cumulative variability plots derived from analysis of the 128 nested sampling sites (Figure 4):

(a) using all 128 observations of D_{50}.
(b) using 64 observations of D_{50} at one time.
(c) using 16 observations of D_{50} at one time.

(after Chorley et al 1966)

225

This application is illustrated by an investigation of contemporary sedimentation on a glacial outwash plain at Ekalugad Fjord (Fig. 7) in east-central Baffin Island. Significant sediment motion occurs during frequent snowmelt floods, jokulhlaups, and heavy summer storm freshets. During high floods as much as 20 to 30 percent of the entire outwash surface may be covered, and changes of stage are abrupt, so that sediment may be entrained and redeposited rapidly with little chance for sorting to occur locally. Flow diverges down sandur, so that velocity and depth decrease, as much larger flow widths and areas are occupied. In such a circumstance, the streams continuously decline in competence downsandur and so deposition of coarse materials under conditions of decreasing competence should dominate the sedimentation pattern. It was hypothesized that mean sediment size and size of coarsest clasts decrease continuously downsandur, though that sediment size decreases away from the mainstream channel in accordance with the flow divergence. The continual loss of the coarsest material from the sediment load should improve sorting downsandur as well. Hence it was further proposed that the variance of sediment size decreased downstream.

Sampling was carried out at 145 points laid out in a square grid pattern on the sandur surface. At each site a sample of 96 cobbles greater than 8 mm b-axis diameter were chosen by randomixed means and axial measurements were made. The results that follow were based on b-axis diameters.

The observed distributions of cobble size and variance on the sandur are shown on figures 8a and 9a. It is obvious that both characteristics do decrease downsandur. Third order trend surfaces were found to provide satisfactory expression of the behavior of both parameters (Figs. 8b and 9b). The analysis of variance of the trend surfaces is given in Table 1. Both grain size and variance were found to decrease down-valley and north across the sandur: i.e. away from South River, which is the primary source of recent sediments, providing confirmation of all three hypotheses.

The residuals from both third order surfaces (fig. 8c and 9c) are generally small, except for large positive values at the proximal end of the sandur and near the distal end on the south side. Both of these areas are just below high cliffs with frequent rockfalls. The large, heterogeneous materials on the surface at these points were derived from rockfall, rather than by glacial-fluvial action. Large negative residuals immediately downvalley in the proximal instance suggest that the large rock sizes found on the surface here have

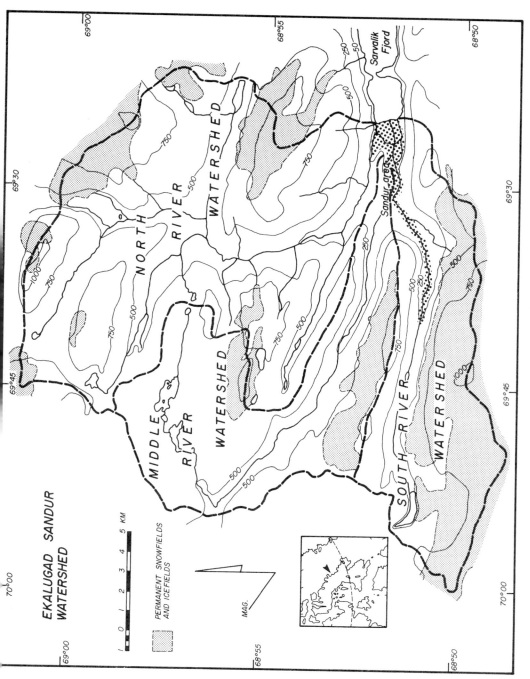

Fig. 7 Location map of Ekalugad sandur watershed, Baffin Island, with sandur area coarsely stippled.

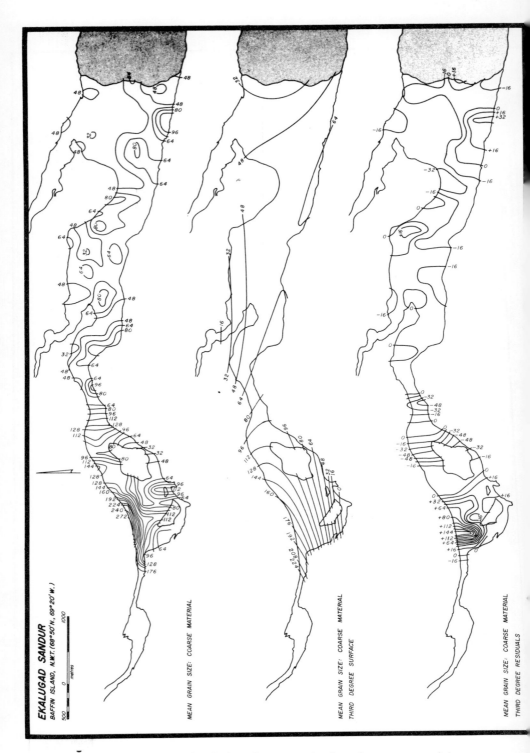

Fig. 8　Baffin Island. Statistics of mean grain size of coarse material on Ekalugad sandur.

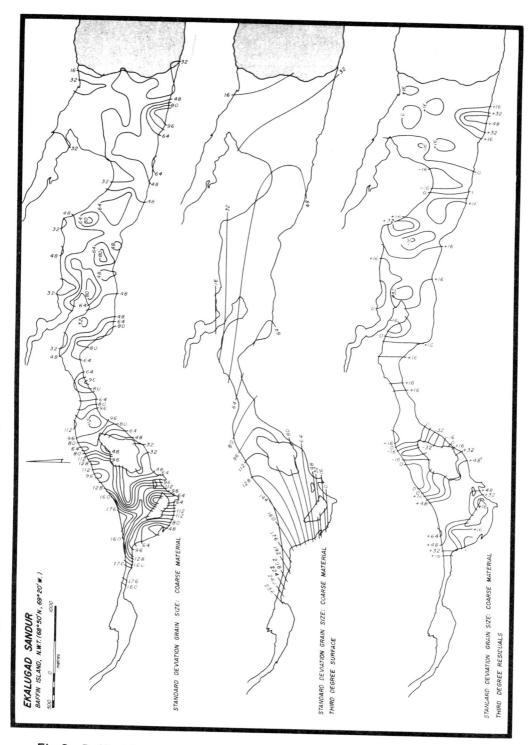

Fig. 9 Baffin Island. Statistics of standard deviation of grain size of coarse material on Ekalugad sandur.

229

TABLE 1

ANALYSIS OF VARIANCE OF TREND SURFACE RESULTS: EKALUGAD FJORD STUDY

Surface	Sum of Squares	df	Mean Square	Residual df	Residual Mean Square	F	R^2
A. Mean size							
1 Increment	120238	2	60119	142	1047	42.72*	0.376
total	120238	2	60119			42.72*	0.376
2	43499	3	14500	139	1125	12.89*	
total	163738	5	32748			29.23*	0.512
3	39453	4	9863	135	866	11.39*	
total	203191	9	22577			26.08*	0.635
Residual	116871						
Total	320062						
B. Standard Deviation Size							
1 Increment	134533	2	67267	142	1048	64.16*	
total	134533	2	67267			64.16*	0.475
2	29751	3	9917	139	857	11.57*	
total	164284	5	32857			38.34*	0.580
3	28795	4	7199	135	669	10.76*	
total	193079	9	21453			32.06*	0.681
Residual	90338						
Total	283416						

* Significant at $\alpha = 0.01$.

distorted the trend fit upward in any case and that the true mean and variance of fluviatile materials that should be expected here is not so great as indicated by the third order trend surface.

In sum, the trend surface patterns and residuals provide an explicable picture of the distributions of coarse material characteristics on Ekalugad sandur, and lend credence to the hypotheses about the physical processes that led to them.

REFERENCES

Agocs, W.B., 1951, "Least squares residual anomaly determination", *Geophysics,* 16: pp. 686-696.

Agterberg, F.P., 1964, "Methods of trend surface analysis", *Colorado School of Mines. Quart.,* 59: pp. 111-130.

, 1970, "Multivariate prediction equations in geology", *Intern. Assoc. Math. Geol. J.,* 2: pp. 319-324.

Agterberg, F.P. and P. Cabilio, 1969, "Two-stage least-squares model for the relationship between mappable geological variables", *Intern. Assoc. Math. Geol. J.,* 1: pp. 137-153.

Allen, P. and W.C. Krumbein, 1962, "Secondary trend components in the top Ashdown pebble bed, a case history", *J. Geol.,* 70: pp. 507-538.

Baden-Powell, D.F.W., 1948, "The chalky boulder clays of Norfolk and Suffolk", *Geol. Mag.,* 85: pp. 279-296.

Beloussov, I.M., N.M. Kozlov, and A.D. Yampol'skiy, 1965, "A new method for the statistical treatment of sounding data", *Oceanology* (in translation), 5 (published 1966): pp. 119-125.

Chayes, F. and Y. Suzuki, 1963, "Geological contours and trend surfaces", *J. Pet.,* 4: pp. 307-312.

Chorley, R.J., D.R. Stoddart, P. Haggett, and H.O. Slaymaker, 1966, "Regional and local components in the areal distribution of surface sand facies in the Breckland, Eastern England", *J. Sed. Pet.,* 36: pp. 209-220.

Cook, A.C., 1969, "Trend-surface analysis of structure and thickness of Bulli seam, Sydney Basin, New South Wales", *Intern. Assoc. Math. Geol. J.,* 1: pp. 53-78.

Grant, F., 1957, "A problem in the analysis of geophysical data", *Geophysics,* 22: pp. 309-344.

Harbaugh, J.W., 1964, "Application of four-variable trend hypersurfaces in oil exploration", in G.A. Parkes, ed., *Computers in the Mineral Industries,* Stanford University, Stanford, California, School of Earth Sciences, *Publications in Geological Science,* 9: 693 pp.

Harbaugh, J.W. and D.F. Merriam, 1968, *Computer Applications in Stratigraphic Analysis,* New York, John Wiley, 282 pp.

Howarth, A.J., 1967, "Trend surface fitting to random data--an experimental test", *Amer. J. Sci.,* 265: pp. 619-625.

James, W.R., 1966, "A double Fourier surface fitting program for irregularly spaced data", Kansas State Geological Survey, *Computer Contributions,* No. 5, 18 pp.

, 1967, "Nonlinear models for trend analysis in geology", in D.F. Merriam and N.C. Cocke, eds., *Computer Applications in the Earth Sciences: Colloquium on Trend Analysis,* Kansas State Geological Survey, *Computer Contributions,* No. 12, pp. 26-30.

Krumbein, W.C., 1956, "Regional and local components in facies maps", *Amer. Assoc. Petrol. Geol. Bull.,* 40: pp. 2163-2194.

,1958, "Measurement and error in regional stratigraphic analysis", *J. Sed. Pet.,* 28: pp. 175-185.

, ,1959, "Trend-surface analysis of contour-type maps with irregular control-point spacing", *J. Geophys. Res.,* 64: pp. 823-834.

,1963, "Confidence intervals on low-order polynomial trend surfaces", *J. Geophys. Res.,* 68: pp. 5869-5878.

,1966, "A comparison of polynomial and Fourier models in map analysis", Geography Branch, ONR, ONR Task No. 388-078, Contract NONR 1228 (36), *Technical Report No. 2,* 45 pp.

Krumbein, W.C. and F.A. Graybill, 1965, "Applications of the general linear model to map analysis", ch. 13 in *An Introduction to Statistical Models in Geology,* New York, McGraw-Hill, pp. 319-357.

Krumbein, W.C. and J. Imbrie, 1963, "Stratigraphic factor maps", *Amer. Assoc. Petrol. Geol. Bull.,* 47: pp. 698-701.

Krumbein, W.C. and H.A. Slack, 1956, "Statistical analysis of low-level radioactivity of Pennsylvanian black fissile shale in Illinois", *Geol. Soc. Amer. Bull.,* 67: 739-762.

Krumbein, W.C. and L.L. Sloss, 1963, *Stratigraphy and Sedimentation,* San Francisco, W.H. Freeman, 2nd ed., 660 pp.

Mandelbaum, H., 1963, "Statistical and geological implications of trend mapping with non-orthogonal polynomials", *J. Geophys. Res.,* 68: pp. 505-519.

McIntyre, D.B., 1967, "Trend-surface analysis of noisy data", in D.F. Merriam and N.C. Cocke, eds., *Computer Applications in the Earth Sciences: Colloquium on Trend Analysis,* Kansas State Geological Survey, *Computer Contributions,* No. 12, pp. 45-56.

Merriam, D.F., 1964, "Use of trend surface residuals in interpreting geological structures", in G.A. Parks, ed., *Computers in the Mineral Industries,* Stanford University, Stanford, California, School of Earth Sciences, *Publications in Geological Science,* 9: pp. 686-692.

Miller, R.L., 1956, "Trend surfaces: their application to analysis and description of environments of sedimentation. 1. The relation of sediment-size parameters to current-wave systems and physiography", *J. Geol.,* 64: pp. 425-446.

Miller, R.L. and E.C. Olson, 1955, "The statistical stability of quantitative properties as a fundamental criterion for the study of environments", *J. Geol.,* 63: pp. 376-387.

Oldham, C.H.G. and D.B. Sutherland, 1955, "Orthogonal polynomials: their use in estimating the regional effect", *Geophysics, 20: pp. 295-306.*

Paul, M.K., 1967, "A method of computing residual anomalies from Bouguer gravity map by applying relaxation techniques", *Geophysics,* 32: pp. 708-719.

Rao, S.V.L.N. and M.S. Rao, 1970 "A study of residual maps in the interpretation of geochemical anomalies", *Intern. Assoc. Math. Geol. J.,* 2: pp. 15-23.

Roberts, M.C. and D.M. Mark, 1970, "The use of trend surfaces in till fabric analysis", *Can. J. Earth Sci.,* 7: pp. 1179-1184.

Romanova, M.A., 1970, Checking linear hypothesis in factor evaluation by use of trend-surface gradients in the investigation of eolian deposits", *Intern. Assoc. Math. Geol. J.,* 2: pp. 231-240.

Simpson, S.M., 1954, "Least squares polynomial fitting to gravitational data and density by digital computers", *Geophysics,* 19: pp. 255-269.

Sneath, P.H.A., 1967, "Trend-surface analysis of transformation grids", *J. Zool.* (London), 151: pp. 65-122.

Svensson, H., 1956, "Method for exact characterizing of denudation surfaces, especially peneplains, as to position in space", *Lund Studies in Geography,* Ser. A. No. 8, 5 pp.

Swartz, C.H., 1954, "Some geometrical properties of residual maps", *Geophysics,* 19: pp. 46-70.

Tinkler, K.J., 1969, "Trend surfaces with low 'explanations': the assessment of their significance", *Amer. J. Sci.,* 267: pp. 114-123.

Thomas, E.N., 1960, "Maps of residuals from regression: their classification and uses in geographic research", State University of Iowa, Department of Geography, *Reports,* No. 2, 60 pp. Reprinted in B.J.L. Berry and D.F. Marble, eds., *Spatial Analysis,* Englewood Cliffs, N.J. Prentice-Hall, 1968, pp. 326-353.

West, R.G. and J.J. Donner, 1956, "The glaciations of East Anglia and the East Midlands: a differentiation based on stone-orientation measurements of the tills", *Geol. Soc. London, Quart. J.,* 112: pp. 69-91.

Whitten, E.H.T., 1959, "Composition trends in a granite: modal variation and ghost stratigraphy in part of the Donegal granite, Eire", *J. Geophys. Res.,* 64: pp. 835-849.

,1960, "Quantitative evidence in palimpsestic ghost stratigraphy from modal analysis of a granitic complex", *Intern. Geol. Congress, XXI. Report,* Part 14, pp. 182-193.

,1961, "Quantitative areal modal analysis of granitic complexes", *Geol. Soc. Amer. Bull.,* 72: pp. 1331-1359.

,1963, "Application of quantitative methods in the geochemical study of granitic massifs", *Roy. Soc. Can. Spec. Pub.* No. 6: pp. 76-123.

,1970, "Orthogonal polynomial trend surfaces for irregularly spaced data", *Intern. Assoc. for Math. Geol. J.,* 2: pp. 141-152.

DISCUSSION: QUATERNARY GEOMORPHOLOGY

Moderator: J. G. Fyles

During each symposium the organisers have attempted to provide an open forum for views on the state of geomorphology with particular reference to Canada. The committee was honoured that Dr. J. G. Fyles of the Geology Survey of Canada was able to act as moderator for the discussion of the 2nd Guelph Symposium on Geomorphology. The condensed form of the discussion below represents the editor's summary of the discussion based on the tape-recording of the session.

Dr. Fyles:

We have heard papers dealing with glacial deposits in the broadest sense. We have also heard papers dealing with frozen ground and these topics together, I think, are the principle elements in the geomorphology of Canada. Both elements are facing us as a major study and I feel that the designers of this symposium have chosen well in focusing on these things. The requirements for such study are exceedingly broad as indicated by the papers which have ranged from experimental investigation on patterned ground through analysis of the thermal regime in the active layer to the varied approaches in the study of glacial and glacial-fluvial deposits which are possible by the application of a variety of quantitative techniques.

I am afraid I do not understand fully the quantification which has developed in our studies over the past few years. Yet mystified or not I am convinced this quantification is fundamentally important if geomorphology is to be a valuable and useful force in the management of the surface of the earth. It is obvious from the discussions we have had today that there are growing pains in the development and application of new techniques. There will be many additional fields of knowledge developed as the people gain familiarity with them. Perhaps we shall sometimes be able to deal with them in the cavalier way in which we have looked at landforms and the materials and processes in the past decades. An interesting thought to me at least in this context is "What is geomorphology at this stage?" We borrow techniques from mathematicians, from chemists and soil mechanics experts and specialists in heat flow. We borrow techniques from hydrologists and all these are brought into the solution of geomorphological problems to form in part a geomorphological point of view. Perhaps even more important questions are, "What are the boundaries and the direction of geomorphology?" and "How should we communicate the consequences of geomorphological investigations and thought to the potential users of our conclusions?"

We are looking at the properties of the materials that form the surface of the earth, the actual form itself, and the processes that are taking place in changing the surface of the earth. I feel that geomorphologists in Canada are

235

now facing a unique challenge to communicate with others because of the tremendous increase in concern with environmental matters, the use of land and the prevention of pollution. All these things have a geomorphological core which relates to the materials at and near the surface of the earth and the processes acting there. If we geomorphologists can communicate our point of view and our conclusions to the users of such information, we will be in a very preferred position indeed. Unfortunately we as geomorphologists do not have a standard clientele. The role of the engineer is well-known the role of the pedologist is well-known; a geologist in mineral resources has his role well-defined. Even the ecologist is being recognized as somebody who speaks loudly and clearly on a particular topic. The position of a geomorphologist differs depending on the situation in which he finds himself, and, in many fields of endeavour people feel that geomorphology is talking to itself.

We have a very real challenge here, a challenge which I find very acute in the Government of Canada today. The Geological Survey of Canada is constantly bombarded with questions about pipelines in the Mackenzie Valley, land-use regulations in the Northwest Territories, a spate of landslides in eastern Canada, dwindling of the mineral economy in Northern Ontario, and plans for major hydro-electric development in the James Bay lowland to name only a few. All of these are immense challenges for geomorphology, if we can pick them up. I feel that we must take up these challenges if the administration, management and development of the surface resources of Canada is to go ahead the way it should.

I trust that in the process of trying to apply our knowledge to these various problems we can use the kind of quantitative techniques which have been developed, and use them effectively, and communicate them in a way that they are understood and used by the planners, engineers, etc., who will have an influence on the use of our land in the coming years.

The recent landslides in Eastern Canada brought home, only too clearly, that despite some decades of investigation of the mechanisms and materials we are still completely unable to predict where a slide may take place. We cannot make a positive suggestion on how to control it or to prevent additional slides in a nearby site. We find that, despite extensive investigation of permafrost and the related landforms, the level of knowledge available to those who are planning for pipeline development in the north is exceedingly low. The Government of Canada is unprepared to undertake a knowledgeable assessment of the situation. Despite the fact that for years we have been saying that glacial drift gives us a useful means of mineral prospecting and geochemists have been saying that they can collect samples and come up with meaningful ways to discover mineral deposits, a careful look at the situation reveals that many things remain unknown, and the record is not very successful.

We certainly must focus on some of these problems and make our purpose that much more meaningful. I would like Bill Shilts to summarise a project which we asked him to do a year ago because we felt that the relationship of knowledge about glacial deposits to bedrock dispersion and geochemistry and its application to the search for mineral deposits was not very well known. We have some results which I hope show you there are a lot of things still to be done.

Editor's Note:
Dr. Shilts then gave a brief paper as an example of applied geomorphology. The paper dealt with the relationship of glacial deposits to bedrock dispersion and geochemistry, and the application of this relationship to the search for mineral deposits.

In presenting his paper Dr. Shilts stated that there were three main points which he wished to stress: (1) Scientists should be addressing themselves to the practical applications of studies of glacial deposits, (2) Studies of drift geochemistry can have pitfalls when undertaken by "the more-or-less uninitiated", (3) In future there will be a preference for research in drift deposits because of the potential applications in prospecting in the large areas covered by glacial deposits.

The study was concerned with trace elements in the glacial deposits of an area to the south-east of the Keewatin Ice Divide. The region was chosen because the history of glacial movement was well-known and the bedrock had been surveyed in detail. Samples of the drift deposits were sieved and analysed for trace elements. The deposits were considered to be either till or re-worked till. As the results were compiled it appeared that the clay-rich deposits showed higher values for trace elements.

Further examination revealed a strong relationship between the concentration of several trace elements and the precentage of clay in the sediment. Concentrations of zinc, copper and lead are directly related to the percentage of clay in the silt-and-clay portion of each sample. Zirconium appears to have an inverse relationship to the clay content. The distribution of the elements being studied seems to indicate a north south pattern. The expected S.E. flow from the ice divide is not obvious. The explanation for this is that the main Wisconsin ice movement was from north to south. The local movement away from the ice divide was only a minor pattern which produced the lineation of surface features at the conclusion of the Wisconsin.

The presentation by Dr. Shilts included data from several sections in the drift deposits and the problem of sampling by auger was underlined by reference to multi-layer sites where augering at a specified depth could have produced samples of any of the several layers. Relationships between ore bodies and the trends of trace elements were also discussed and the heavy mineral data were compared with the trace element information. Dr. Shilts concluded by emphasizing the importance of a full understanding of data for applications in prospecting. There is much information still to be derived from the data already collected in the study of drift deposits.

Following this presentation, Dr. Fyles acted as chairman for an open discussion. *Brookes (York)* asked Dr. Fyles to provide some information concerning the activities of GSC in organizing a central data bank, to facilitate an exchange of data among geomorphologists in Canada. Dr. Fyles replied that while GSC has a data filing system, it has, so far, been ineffective in dealing with certain problems. They are presently operating on a "part-way basis" until an alternative method is found. *Lewis (CCIW)* commented that the problem with such data file systems is the difficulty encountered in entering your own data in a form which is useful to you, but at the same time useful to others. He noted that this would encourage a standardization in reporting data.

Sutterlin (UWO) added that there is a national sub-committee on computer applications in the geological sciences. This committee is examining problems such as those pointed out above.

It was pointed out by Dr. Fyles that industry and government agencies, as well as academic institutions, are interested in having access to data concerning unconsolidated landforms and land-use, through a data file system. The problem again is how to input this information to everyone's satisfaction.

The theme of outside interest in geomorphological investigations was expanded by *Corte (Argentina).* He presented an example of civil engineers showing an interest in freeze-thaw studies, to aid them in constructing pipelines across a land surface undergoing this cycle.

Bird (McGill) turned the discussion back to Dr. Fyles' summary remarks in regard to the failure of geomorphologists to answer important questions. He put forward the hypothesis that it is the politician who is at fault for asking the wrong questions in the first place. He calls for an education of "not only the geomorphologists who are involved in this, but of the people who ask the questions." Dr. Fyles concurred with these remarks, and cited examples from his own experience.

Andrews (Colorado) decried the fact that, in dealing with practical applications of geomorphology, we are always forced to operate on an "ad hoc" basis. We are consulted only after the decision (e.g. to build a pipeline), has been made.

A further example of this request for an "instant decision" was provided by *Packer (UWO),* and dealt with the decision by London City Council to build a bridge. Only after the site of the construction, and the date for its completion had been decided, were the geomorphologist/geologists consulted.

Packer (UWO) also referred back to the remarks made earlier concerning the need for geomorphologists to sell themselves to the potential users of their investigations. He pointed out that an organisation called the "Institute of Ontario Geomorphologists" was already examining this problem closely.

Dreimanis (UWO) commended the report which Dr. Shilts had presented, and called for more emphasis on such basic research.

Harris (Calgary) called for a change in the education of Canadian graduate students in geomorphology. He felt that the emphasis should be directed toward "problem solving", and away from the traditional types of research.

Dr. Fyles concluded the discussion with the comment that, while not trained in a "problem oriented" system, geomorphologists have shown the ability to apply their knowledge to practical problems. They have also shown a facility for relating well to people in allied fields, due to the broad spectrum of subjects studied by these students of geomorphology.

ADAPTATION OF THE CONTINUOUS PARTICLE ELECTROPHORESIS SYSTEM FOR THE QUANTITATIVE ANALYSIS OF CLAY MINERALS

L. C. Hodgson and J. B. Reynolds

In the study of clay minerals one of the most persistent problems is the need for a more accurate method for quantitative analysis. At present, the most widely used method is quantitative X-ray diffraction analysis. This method is based on the fact that peak intensity on an X-ray diffraction pattern is a function of the amount of clay present (Weaver, 1958). There are, however, so many other factors influencing peak intensity that analyses based on this method are reduced to a semi-quantitative level (Brown, 1961; Johns, Grim and Bradley, 1954; McNeal, 1968; Pierce and Seigel, 1969).

A relatively new method of clay mineral analysis shows a great deal of potential for helping to alleviate this problem. This "new" method is the electrophoretic separation of clay minerals. Electrophoresis is defined as the movement of charged colloidal particles under the influence of a given electrical field (Drever, 1969). Clay minerals fit the category of charged particles, with the charge resulting mainly from isomorphous substitution within the crystal lattice. However, if everything smaller than 2 microns is considered to be a clay, many of the particles are obviously not going to be of colloidal size. This was one of the major problems that beset the first researchers who attempted electrophoretic separation of clay minerals.

This problem was overcome, however, in 1966 when an electrophoresis cell was designed for the continuous microfractionation of particles larger than colloidal size (Strickler, Kaplan and Vigh, 1966). This cell, known as the Continuous Particle Electrophoresis System, or CPE System for easier reference, was originally meant to be used in biomedical research, but in 1969 two separate studies were conducted using this CPE system to separate mixtures of clay minerals. Both researchers reported success in qualitatively separating mixtures of clay minerals, and both recommended that further investigation of this potentially valuable apparatus was warranted (Drever, 1969; Park, 1969).

It was this recommendation that prompted the beginning of investigations of the feasability of using this CPE System for a quantitative analysis of a clay mineral mixture. Soon after research began, however, it became obvious that many modifications would have to be made before it could be of value in a quantitative study. It is these modifications which form the subject matter of this paper.

The Beckman CPE System possesses several features which make it a valuable piece of equipment for use in clay mineral separations. In the original system the supporting medium, which is an electrolyte at a low concentration, enters at the top and back of the cell (see Figure 1), and flows in a two millimeter thick curtain at a speed regulated by the "curtain flow meter". The clay sample is suspended in the same electrolyte and is fed into the centre of this free flowing curtain, below the curtain buffer inlet, by a length of fine polythene tubing. The sample and the electrolyte flow down between the two electrodes (see Figure 1), exit via 48 exit tubes and are collected in 25 ml. test tubes. One additional feature of this system is that the electrodes are continually rinsed to dissipate heat and neutralize electrolysis products.

In order to carry out a quantitative analysis using this apparatus it is necessary to operate it for 8 to 10 hours at a time. This is necessary due to the relatively large amount of sample required, and the small amount that can be passed through the machine at one time. The need for an extended running time led to the first modifications of the CPE System.

The build up of trapped particles on the filter in the curtain buffer system increased the "resistance" of the filter to such an extent that the original curtain buffer pump, with its 1/10 HP size, was unable to maintain a constant flow rate for the duration of the experiment. In addition, this pump was too weak to force air locks out of the system without being primed. For these reasons it was replaced by a 1/4 HP variable speed motor (see Figure 2).

The second problem encountered due to the length of running time was the build up of heat in the electrode rinse system. Experiments were run at a voltage gradient of 75 v/cm, for eight hours. The heat generated during this time caused convection currents in the curtain buffer, which in turn, interfered with migration of the clay. To overcome this problem the electrode rinse reservoir was attached to the outside of the side door, and a plastic coil was placed inside the rinse bottle. Cold water, at a constant pressure is circulated through the coil. With this modification the problem of heat build up in the electrode rinse system was effectively eliminated.

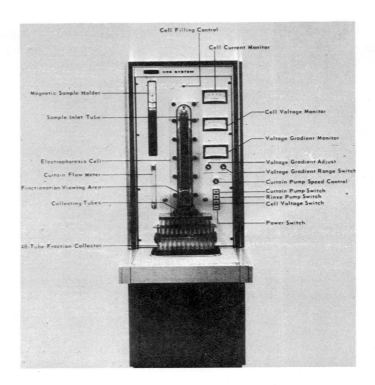

Fig. 1 Front panel of Beckman CPE System (Source: Beckman CPE Operators' Manual).

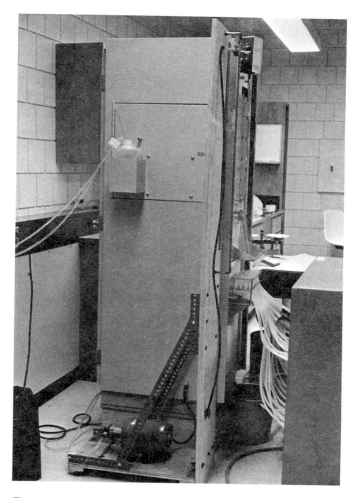

Fig. 2 Modifications introduced to the CPE System.

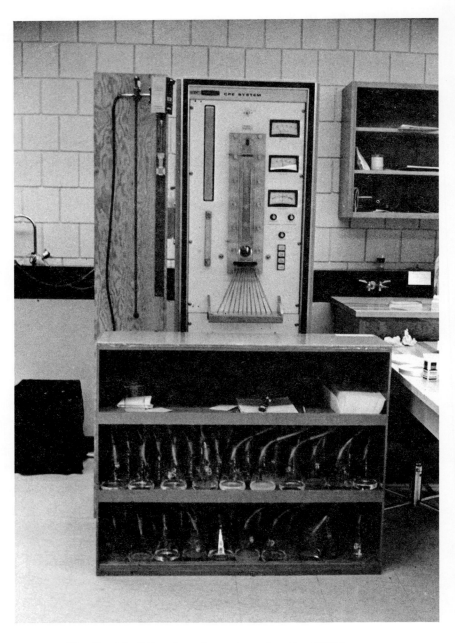

Fig. 3 Modifications introduced to the CPE System.

The remaining modifications were made necessary, not by the long running time, but by the increased quantity of sample required for a quantitative analysis, as opposed to a qualitative one. The concentration of clay in the sample holder is 10X greater for a quantitative analysis. A Corning Stirrer was utilized to keep the sample agitated. Due to excess vibration it had to be mounted on a plywood base adjacent to the CPE Cabinet (see Figure 2).

The last modification was to the sample collection system. In the original CPE system, sample and electrolyte were collected in 25ml. test tubes. With a curtain flow rate of 25 ml./min. these test tubes had to be changed every hour. To overcome this problem the test tubes were re-placed by 750 ml. beakers. A cabinet was built to hold the beakers and a polythene funnel inserted in the end of a length of polythene tubing ensured that the drops from the tubes are transferred to the correct beaker (see Figure 3).

Experiments have been carried out using the modified CPE System in an attempt to separate a 1:1:1 mixture of illite, montmorillonite and kaolinite, and also a Leda clay. In the first case there was a separation of illite from one fraction and montmorillonite from another at pH 7.5 with a 10^{-4}M $NaHCO_3$ electrolyte. Under the same test conditions an amorphous material was separated from the Leda clay.

Although a complete separation has not yet been obtained, the potential value of this new method warrants further research in this field.

REFERENCES

Brown, George (ed.), (1961), X-Ray Identification and Crystal
    Structure of Clay Minerals, London Mineralogical Society,
    London.

Drever, James, (1969), "The Separation of Clay Minerals by Continuous
    Particle Electrophoresis". Dept. of Geological and Geophysical
    Sciences, Princeton University, Princeton, N.J., Technical Report
    # 1.

Johns, W. D., Grim, R., Bradley, W. F., (1954), "Quantitative
    Estimation of Clay Minerals by Diffraction Methods". Journal of
    Sed. Petrology, Vol. 24, No. 4, pp. 242-251.

McNeal, Brian L., (1968), "Limitations of Quantitative Soil Clay Miner-
    alogy". Soil Science Society of America Proc., Vol. 32, pp.
    119-121.

Park, R. G., (1969), The Electrophoretic Separation of Mixtures of Pure
    Clays (University of Idaho, Ph.D.) University Microfilms, Inc.,
    Ann Arbor, Michigan.

Pierce, J.W., Siegel, F. R., (1969), "Quantification in Clay Mineral
    Studies of Sediments and Sedimentary Rocks". Journal of Sed.
    Petrology, Vol. 39, No. 1, pp. 187-193.

Strickler, A., Kaplan, A., Vigh, E., (1966), "Continuous Microfraction-
    ation of Particle Mixtures by Electrophoresis". Microchemical
    Journal, Vol. 10, pp. 529-544.

Weaver, C., (1958), "Geological Interpretation of Argillaceous Sedi-
    ments. Part 1. Origin and Significance of Clay Minerals in
    Sedimentary Rocks". Bulletin of the American Assoc. of Petroleum
    Geologists, Vol. 42, No. 2, pp. 254-271.

DATING CAVE CALCITE DEPOSITS BY THE URANIUM DISEQUILIBRIUM METHOD: SOME PRELIMINARY RESULTS FROM CROWSNEST PASS, ALBERTA.

D. C. Ford, P. Thompson and H. P. Schwarcz.

PRINCIPLES OF URANIUM — THORIUM RADIOMETRIC DATING METHODS.

Uranium is present in trace amounts in many rocks. When these are chemically weathered it may be leached as a solute. Should host water contain other, more common, solutes in such abundance that there is supersaturation and precipitation occurs, some of the uranium will be co-precipitated. Trapped within the new deposit it is a potential clock, recording the time elapsed since precipitation by measurable radioactive decay.

The uranium radioactive decay scheme yields four possible methods of dating a host precipitate:—

1) the isotope ratio, U—234/U—238. Potentially this is effective within the timespan, 50,000 — 1,500,000 years B.P.
2) the isotope ratio, Th—230/U—234. The dating timespan is 2,000 — 350,000 years B.P.
3) the isotope ratio, Th—230/Pa—231. The dating timespan is 2,000 — 200,000 years B.P.
4) the isotope ratio, Ra—226/U—234. The dating timespan is 300—7,000 years B.P.

The first method rests upon the fact that in all ocean waters U—234 and U—238 are present in the ratio 1.15/1.00, (Thurber 1962; Koide and Goldberg 1965). In a precipitate the U—234 excess decays to unity by alpha-particle emission at a rate fixed by the half-life of the isotope. With present-day counting devices this effect is measurable to approximately six half-lives of U^{234} (c. 1.5 million years). The formal expression of the relationship is:—

$$\left[\left(\frac{U^{234}}{U^{238}} \right)_t - 1 \right] = \left[\left(\frac{U^{234}}{U^{238}} \right)_o - 1 \right] e^{-\lambda t}$$

(1)

where $\left(\dfrac{U^{234}}{U^{238}} \right)_t$ is the ratio at time "t", (measured ratio).

$\left(\dfrac{U^{234}}{U^{238}} \right)_o$ is the ratio at time zero, (initial ratio)

and λ is a constant, $(0.693/t^{1}/_2)$.

"t" is the age, (time elapsed since deposition).

The basis of the second method is the contention that no thorium is deposited in precipitates from natural aqueous solutions, (Barnes et al. 1956, Fornaca-Rinaldi 1968, Duplessy et al. 1970). Any Th−230 that is measured in the extract of a precipitated rock therefore must have been produced by decay of U−234 since deposition of the solid phase. The age of a measured deposit is determined from the equation:−

$$\left[\frac{Th^{230}}{U^{234}}\right] = \frac{(1 - e^{-\lambda 230 t})}{(U^{234}/U^{238})_t} + \left[1 - \frac{1}{(U^{234}/U^{238})_t}\right] \cdot \left[\frac{\lambda_{230}}{\lambda_{230} - \lambda_{234}}\right] \cdot$$

$$\cdot \left[1 - e^{-(\lambda_{230} - \lambda_{234})_t}\right] \qquad (2)$$

where Th_{230} is the ratio at time "t" (measured ratio) and λ_{230}, λ_{234} are decay constants of Th-230, U-234, respectively $0.693/t_{1/2}$ (thorium) and $0.693 t_{1/2}$(uranium).

The third method, measurement of the ratio of thorium−230 to protacti nium−231, (produced by α-decay of U−235, the second primary uranium isotope) is applicable mostly to deep-sea cores and has not been pursued by the present authors, (see Rosholt et al. 1961, Rona and Emiliani 1969). The fourth method, radium−226/uranium−234, measures within the timespan covered by the long-established C−14 method and so is of little interest.

In addition, radium is an alkaline earth element. Therefore, it is to be expected th it will be included in initial precipitations of other alkaline earths, such as the cav calcite considered below. The amount of radium so deposited might not be readil determined. Measurement of the Ra−226/U−234 ratio is useful for checking "recent" additions or removals by leaching of radioisotopes in deposits of interest

INVESTIGATIONS AT McMASTER UNIVERSITY.

There are comparatively few published results of dating precipitates by the uranium methods. Most are concerned with marine and lacustrine deposits, (Thurber et al. 1965, Kaufman and Broecker 1965). In theory, three of the methods are equally applicable to the precipitates from terrestrial waters. I 1967 the authors commenced an investigation of cave calcite precipitates − the familiar stalactite, stalagmite and flowstone deposits of limestone caves Thompson took charge of detailed work. He has established two important points:−

(a) although the ratio, U−234/U−238, is 1.15 [1.00] in the oceans, it varie from 1.0 to 3.0 [1.0] in fresh karst waters. Variation may occur in space a time. This severely limits application of the first uranium method with its exciting time range of 1.5 million years.

(b) thorium−230 in cave calcite deposits is usually authigenic i.e. the syste is closed to any external contamination immediately the calcite and uraniu are co-precipitated. The presence of *any* initial Th−230 can be monitored b the absence of Th—232, a long-lived isotope which is always associated with detrital Th—230. The Th—230 : Th—232 activity ratio is a measure of contamination by detrital thorium. Preferably, the ratio should be greater than 50. Low ratios indicate that the measured age is slightly too high where the Th—230/U—234 method is used. It is not possible to correct for this error.

The majority of cave calcite samples processed to date at McMaster University derive from the karst of W. Virginia, which has been used as readily accessible type-area for the study. But a few from the Crowsnest Pass region in the Canadian Rockies have also been measured. The remainder of this paper comments upon the latter to illustrate the potential power of the uranium method and its karst application in resolving major problems of landform evolution in the later Pleistocene.

LIMESTONE CAVES AND PLEISTOCENE LANDFORM STUDIES IN CANADA

Most or all of Canada was glacierized during the later glacials. In any region the dominant surface landforms are attributable to the later stages of the last glacial incursion in that region or they are post-glacial. Information upon the magnitude, rates, base levels, etc. of earlier erosion is restricted to a few highly-localized, time-limited, (and lucky) cores, foundation excavations, etc. and a handful of controversial mountain-top refugia. For a number of years one of the authors (Ford), has contended that the best opportunity of probing events of regional scale before the obliterating effects of the last glaciation is offered by caverns in limestone regions. Caverns are a different sort of refugia and contain calcite deposits which have now been shown to be dateable.

The argument is illustrated in Figure 1. Figure 1A shows the effects of repeated glacial erosion, (three glacierizations), of a hypothetical plateau landscape. Relief first increases, then a broadly equilibrium form is attained. An observer of surficial forms and deposits at Time 2, (after a first glacierization), has very little wherewith to reconstruct the profile of Time 1: similarly with Time 2 observed from a Time 3 position, etc. Reconstruction of any profile or merely base level that is older than the previous inter-glacial is quite infeasible. At Time 4 all of the rock mass at and above an arbitrary datum in the profile of Time 1 has been removed.

In Fig. 1B a simple cave of the deep phreatic type, (loosely, a "sub-water-table" cave — see Ford 1965, 1971a), is inserted, generated during Time 1 by a groundwater stream draining to the righthand side of the illustration. In all later profiles it is a fossil, the groundwater source having been diverted or quite erased. But portions of the cave survive until the enclosing rock mass is entirely destroyed. At Time 4 they are the only vestige that remains of conditions at Time 1.

In many fossil caverns there are two distinct depositional facies — those of the entrance zone and those of the interior. All readers will be aware of the

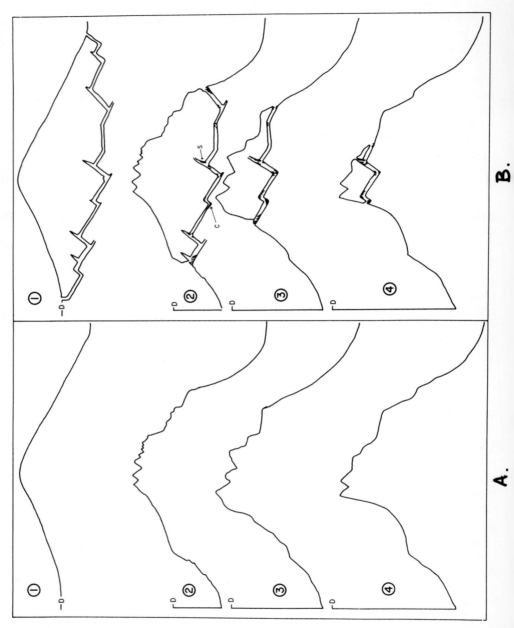

Fig. 1 Successive cross-profiles of a hypothetical highland subjected to three alpine glacierizations. 1A — the common situation. 1B — a cave system is present and retained as a fossil during later glacierizations.
D = an arbitrary height datum which is constant.
c = clastic sediment trap in the cave.
s = stalactite, etc. deposition.

immense amount of information on the Pleistocene that has been obtained from entrance facies in different parts of the world. But in the Canadian context of intensive, repeated glacierizations, entrance facies are nearly as vulnerable as the forms and deposits of the surface profile. Cave entrance facies extend only a few tens or, at most, hundreds of feet into a cave. These parts are removed by glacial erosion, as Figure 1B illustrates. It appears a pipe-dream to suppose, as some have, that cave habitats of pre-Wisconsin man will be found in those parts of Canada over-run by Wisconsin ice.

Only the interior cave facies will survive in an ancient cavern in glaciated regions, (with recent entrance facies being superimposed in the current entrance zones.) The interior facies have three common components, which may be found in any stratigraphic order and are often repeated. These are a, local limestone blocks fallen from roof and walls; b, fluvial clastic deposits from external sources; c, cave calcite deposits. Organic material of any kind is extremely rare. It is the great merit of the uranium method that, for the first time, it permits a common component of the interior facies to be dated. An investigator at Time 4 of Figure 1B can reconstruct some of the erosional circumstances of Time 1 and place at least an upper limiting date upon them.

PRELIMINARY RESULTS FROM CROWSNEST PASS, ALBERTA.

Figure 1 was not entirely hypothetical. The initial and final profiles are broad representations of the High Rock Range of the Rockies in the vicinity of Crowsnest Pass. The Range strikes N. — S. It is composed predominantly of massive Devonian and Mississipian limestones. It is deeply dissected by cirques. During the Wisconsin glacial a valley glacier flowed eastward through the Pass. Its trimline is at approximately 6,700 feet m.s.l. The Pass floor is at 4,400 feet m.s.l. and summits rise to 8,500 — 9,000 feet: thus, there is some 4,000 — 4,500 feet of local relief in this part of the Canadian Rockies.

Since 1968 McMaster University parties have located and explored many limestone caverns in this part of the Range. Figure 2 locates those that are known on the East face: there are equally as many on the West. Most are phreatic and fossil, fragments of extensive conduit systems that have been dissected by valley and cirque entrenchment and cliff recession. The fragments indicate that palaeo-groundwater flow was towards the Pass, which is evidently a long-established feature. There is also modern groundwater flow to the Pass, (Ford 1971b), indicating that ancient patterns of karst water organisation are maintained down to present times, albeit at lower elevation.

The altitudes of the accessible caverns range from 4,440 feet to above 8,500 feet m.s.l., i.e. they are distributed throughout the local relief of the High Rock Range. Many of the fossil caverns contain the remains of large, ornate stalactites, stalagmites and flowstones. These are not actively enlarging today: many have been shattered by earth movements or frost action. Modern calcite deposition is limited to the growth of small stalactites in the lower caves.

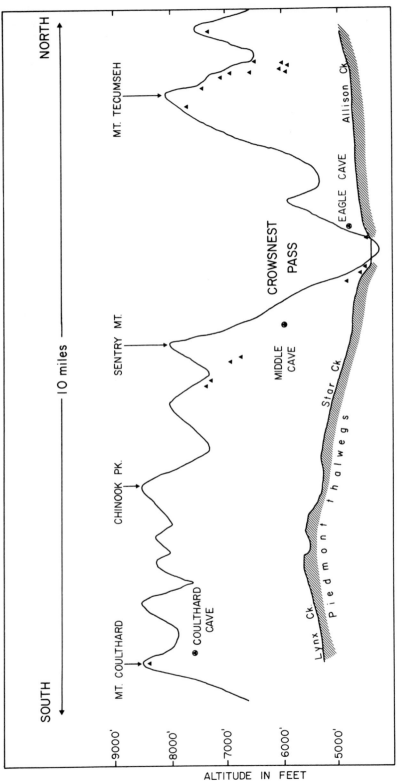

Fig. 2 Sketch profile of the East face of the High Rock Range about Crowsnest Pass, Alberta – B.C. Triangles indicate the entrance of known fossil cave systems. Th. – 230/U-234 dates from Coulthard Cave, Middle Cave and Eagle Cave are discussed in the text.

Specimens of ancient or recent calcite have been taken from eight caves and the modern uranium ion concentrations have been measured in underground stream and percolation waters. At time of writing, only three of the calcite samples have been analysed. They derive from the three caves named on Figure 2 and were chosen for a preliminary analysis because of the wide range of altitude at which they were deposited. Measurement was by the Th−230 : U−234 method. Data are quoted in Table 1.

All three of the calcite specimens were deposited in phreatic cave sites which were air-filled when net deposition began, i.e. the caverns were drained and so stood above the springs, which would normally be at the level of a valley floor. The sites of the Mt. Coulthard and Middle Cave specimens are such that these caves must have been permanently drained before the observed deposition. Eagle Cave could have been periodically activated by floods when net deposition commenced. Middle Cave and Eagle Cave certainly discharged to the Pass. Therefore, the floor of the Pass was below 5,900 feet, (modern altitude), 275,000 years ago and below 4,750 feet, (modern altitude), 200,000 years ago. These altitudes for the floor are maxima and their ages minima. Mt. Coulthard Cave may have discharged into a proto-Star Creek basin, (see Figure 2), with a thalweg at much higher elevation than the Pass but the accordance of its dates with the oldest of Middle Cave, which is 1,550 feet lower in altitude, suggests that it was hanging high above local valley floors when the recorded calcite deposition began. At least 2,600 feet (or 63%) of the present relief about the Pass existed 275,000 years ago. 90% of the relief existed at 200,000 years B.P., before the onset of the penultimate glacial. Since that time the floor of the Pass has been lowered by no more than 400 feet. The mean rate of lowering for the past 200,000 years is 2 feet per thousand years or somewhat less. Given that it is averaged across at least two periods of glacial erosion and a long span of fluvial action during the last interglacial, the mean rate is not a very significant figure. It is noteworthy, however, that it is much smaller than others have obtained by extrapolating modern measures of erosion elsewhere in the Canadian Rockies, (H. McPherson, University of Alberta; pers. comm.).

It is recognised today that, probably, there have been many more glacials than the four that are well-attested in eastern and central North America, (Broecker and Van Donk 1970). This will be especially true of a glacier source area such as the Canadian Rockies. It appears that, about Crowsnest Pass, the mountains are old features, largely attributable to the earlier glacials and interglacials of which no overt traces remain. With such relief at the Pass, it may be supposed that the High Rock Range to N. and S. of it and the Livingstone Range, (10 miles to the East and downstream), also possessed much of their modern relief 200,000 − 300,000 years ago.

It is stressed that these conclusions are preliminary. Further dates from cave calcite samples taken in other caves on the East and West flanks of the High Rock Range should build up a surer picture of the maximum valley floor altitudes and minimum relief of the past 400,000 years. In addition, it is hoped that they will establish distinct phases of calcite growth or inactivity which will be indicative of major climatic changes.

TABLE 1

RADIOMETRIC DATA AND TH-230; U-234 AGES OF CAVE CALCITE DEPOSITS, HIGH ROCK RANGE AT CROWSNEST PASS, ALBERTA.

Site	Material	U conc. (p.p.m.)	$\left[\dfrac{U\text{-}234}{U\text{-}238}\right]_t$	$\left[\dfrac{U\text{-}234}{U\text{-}238}\right]_o$ +	$\dfrac{Th\text{-}230}{U\text{-}234}$	$\dfrac{Th\text{-}230}{Th\text{-}232}$	Th-230 age (yrs. B.P.)
1. *Mt. Coulthard Cave* altitude = 7,520 ft. m.s.l.	broken stalagmite						
	(a) core	0.23	1.07 ±0.023*	1.16 ±0.062	0.92 ±0.029	162	296,000 ±33,000
	(b) outer sheath	0.25	1.00 ±0.016	1.01 ±0.029	0.86 ±0.025	99	235,000 ±19,000
2. *Middle cave of Sentry Mtn.* altitude = 5,975 ft. m.s.l.	broken flowstone ledge deposit (6" thick)						
	(a) base	0.12	0.97 ±0.026	0.93 ±0.045	0.87 ±0.038	18	273,000 ±37,000
	(b) centre	0.07	1.07 ±0.035	1.11 ±0.056	0.75 ±0.032	24	160,000 ±13,000
	(c) upper	0.07	1.10 ±0.028	1.14 ±0.038	0.58 ±0.025	10	99,000 ± 6,000
3. *Eagle Cave* altitude = 4,750 ft. m.s.l.	broken flowstone ledge deposit, 2" inches thick.	0.28	1.40 ±0.023	1.70 ±0.064	0.87 ±0.026	46	198,000 ±13,000
4. Modern water of the area — the *Crowsnest Spring.* altitude = 4,450 ft. m.s.l.	Average of 3 determinations	—	2.1 ±0.30	—	zero		

+ calculated using the equation
$$\left[\frac{U\text{-}234}{U\text{-}238}\right]_o = \left\{\left[\frac{U\text{-}234}{U\text{-}238}\right]_t - 1\right\} e^{\lambda t} + 1 \text{ and the Th-230 age for "t"}$$

* the errors quoted are one std. deviation.

ACKNOWLEDGMENTS

This research was supported by a Special Equipment Grant and Operating grants from the National Research Council of Canada and by the Science Division, McMaster University. Their assistance is acknowledged most gratefully.

REFERENCES

Barnes, J. W., E. J. Lang and H. A. Potratz 1956. "Ratio of ionium to uranium in coral limestones." Science 124, 175.

Broecker, W. S. and J. Van Donk 1970. "Insolation changes, Ice volumes and the O^{18} record in deep sea cores." Rev. Geophysics and Space Physics, 8, pp. 169-198.

Duplessey, J. C., J. Labeyrie, C. Lalou and H. V. Nguyen 1970. "Continental climatic variations between 130,000 and 90,000 years B.P." Nature, 226, pp. 631-633.

Ford, D. C. 1965. "The origin of limestone caverns: a model from the central Mendip Hills, England." Natl. Speleo. Soc. America, Bull. 27 (4), pp. 109-132.

Ford, D. C. 1971a. "Geologic structure and a new explanation of limestone cavern genesis." Trans. Cave Res. Gp., G.B. 13(2), pp. 81-94.

Ford, D. C. 1971b. "Characteristics of Limestone Solution in the Southern Rocky Mountains and Selkirk Mountains, Alberta & British Columbia." Can. J. Earth Sciences, Vol. 8(6), pp. 585-609.

Fornaca-Rinaldi, G. 1968. "$230Th/234Th$ dating of cave concretions." Earth and Planetary Sci. Letters, 5, pp. 120-122.

Kaufman, A. and W. S. Broecker. 1965. "Comparison of Th230 and C14 ages for carbonate materials from lakes Lahontan and Bonneville." Jnl., Geophys. Res. 70, pp. 4039-4054.

Koide, M. and E. D. Goldberg. 1965. "U–234/U–238 ratios in seawater." Progress in Oceanography, 3, pp. 173-177. Pergamon Press.

Rona, E. and C. Emiliani. 1969. "Absolute dating of Caribbean cores P6304-8 and P6304-9." Science, 163, pp. 66-68.

Rosholt, J. N., C. Emiliani, J. Geiss, F. F. Koczy and P. J. Wangersky. 1961. "Absolute dating of deep-sea cores by the Pa–231/Th–230 method." Jnl. Geol. 69 pp. 162-183.

Thurber, D. L. 1962. "Anomalous U–234/U–238 in nature." Jnl. Geophys. Res., 67, p. 4518.

Thurber, D. L., W. S. Broecker, R. L. Blanchard and H. A. Potratz, 1965. "Uranium series ages of Pacific atoll coral." Science 149, pp. 55-58.

CARTOGRAPHIE GÉOMORPHOLOGIQUE DÉTAILLÉE EN MORPHOLOGIE GLACIAIRE SECTEUR DE BURY, COMTÉ DE COMPTON, QUÉBEC

Par J. M. M. Dubois

ABSTRACT

The Bury brook watershed is located in the Ordovician slates on the south-eastern side of the Saint Francis valley, south-east of the Stoke anticline and in the center of the Connecticut-Gaspe Siluro-Devonian synclinorium.
The relief pattern is probably controlled by the development of an ortho-gonal system of joints. The last glacial recession left a bedrock controlled topography. The interfluves are mainly composed of silty boulder till veneers, whereas the valleys and depressions evolved in glacial stagnation zones, non-related to an ice front; the downwasting topography of these zones was controlled by the bedrock irregularities. The grain-size distribution of the till is directly dependent on the altitude, whereas the grain-size variations of the fluvioglacial deposits are essentially dependent on the obturation conditions. During the downwasting, meltwater flowed towards the body of ice or its fringes but in the same direction as the present stream, except at the begining when southern outlets existed, due to the ice dammin of the valley. The detailed geomorphological mapping in color (1:25 000) of the area has permitted us to determine the downwasting stages.

Le bassin du ruisseau de Bury se situe dans les schistes ordoviciens du versan sud-est de la vallée du Saint-François, près de la barre structurale de Stoke, e au centre du synclinorium siluro-dévonien de Connecticut-Gaspé (Cady, 1960). Le modelé de relief y serait commandé par le développement d'un système orthogonal de fractures dont rend compte les orientations des éléments du relief. Ces fractures liées à une tectonique tardive auraient été exploitées par les cours d'eau antécédents développant des tributaires qui ajustaient progressivement le réseau hydrographique à la structure (Bird, 1970). Ceci se remarque par le fait que certaines vallées ne sont plus du calibre des cours d'eau actuels et qu'il y a évidence de quelques captures. Par exemple il est fort probable que la rivière au Saumon se déversait dans la Saint-François par l'intermédiaire des ruisseaux Bown et de Bury. La dernière récession glaciaire qui date de 12,000 à 13,500 ans B.P. selon B.C. McDonald (1967) y a laissé une topographie contrôlée par la roche en place

1 — CARTOGRAPHIE:

La **chronologie** employée ici n'est que relative vu la rapidité d'évolution des phénomènes en période glaciaire, et post-glaciaire, vu le fait qu'une forme finiglaciaire à un endroit peut être contemporaine d'une forme glaciaire un peu plus loin (Ritchot, 1963) et vu les lacunes de datation absolue dans la région. La succession des épisodes morphogénétiques commande cette chronologie.

A l'instar de M. Klimaszewski (1963), et au Québec, de G. Ritchot (1963) et de A. Poulin (1968), une couleur a été attribuée à chacun des épisodes afin de les mieux différencier sur la carte. Les épisodes représentés ici sont: le Wisconsin en violet, le Finiglaciaire d'accumulation en vert, le Finiglaciaire d'affaissement en orange, l'Holocène en brun foncé dont l'Actuel en noir. Quant aux formes de roche en place, aucune chronologie n'a été déterminée. Cependant les affleurements ont été indiqués au moyen de traits rouges conformes aux directions de schistosité.

La **morphographie** des formes est nécessairement étroitement liée à leur genèse de sorte qu'un figuré peut représenter un même phénomène à différents épisodes selon la couleur qui lui a été attribuée; il en est de même pour le faciès des dépôts pléistocènes. Le faciès d'un dépôt peut se déterminer par l'agencement de plusieurs des éléments figurés comme suit: blocs anguleux, blocs émoussés et galets, sable, limon et argile. Le faciès du Wisconsin est quelque peu différent, je n'ai distingué que la moraine à gros blocs.

2 — ANALYSE DE LA CARTE:

L'analyse de la carte ainsi conçue démontre d'une façon générale que les interfluves sont le domaine du Wisconsin, les têtes de vallons et les fonds de vallées le domaine du Finiglaciaire, tandis que l'Holocène est contigü aux talwegs les plus importants.

Dans le domaine du Wisconsin d'abord, la moraine de fond généralement peu épaisse est uniformément sous-jacente aux dépôts qui lui sont postérieurs. De la même façon, cette moraine de fond recouvre les formes de la roche en place, même celles qui ont des pentes de plus de 16° et elle est plus épaisse et moins grossière vers les vallées et dans les

BASSIN DE BURY
Géomorphologie du secteur amont

0   1 mille

LÉGENDE

Barres de roche en place liées aux directions de schistosités et roche moutonnée

Vallon polygénique et banquette de roche en place masquée de moraine

Epandage fluvioglaciaire

Butte de fusion (kame)

Bourrelet de fusion

Talus de terrasse d'obturation (terrasse de kame)

Cône de déjection holocène

Talus d'érosion holocène

Talus d'érosion actuel

Ravinement holocène en U et en V

Ravinement actuel en U et en V

Cour d'eau et plan d'eau

Dépôts organiques de marécage

Alluvions holocènes

Moraine limoneuse à gros blocs

Moraine limoneuse

Moraine d'ablation

Placages sableux sur moraine

Sables fluvioglaciaires

Gravier fluvioglaciaire

fonds topographiques que sur les interfluves et les pentes importantes. C'est donc dire que les placages morainiques vers les sommets d'inter-fluves, en plus d'être truffés de gros blocs, ont une matrice plus grossière que la moraine limoneuse épaisse des bas de versants. Il y a une relation directe aussi entre l'importance de la pente et la grossièreté de la matrice. Des comptages ont démontré que le matériel grossier des moraines est à plus de 95% d'origine locale. De plus, le fait que l'émoussé des sables grossiers (0.5 - 2mm) de la moraine augmente en raison de l'altitude peut laisser supposer que sur les sommets les matériaux, tout en étant locaux, devaient venir d'un peu plus loin que ceux d'en contrebas.

Contrairement à la moraine de fond, les dépôts fluvioglaciaires sont de façon générale plus fins en altitude et moins émoussés. Le matériel est cependant d'origine locale aussi.

La plus importante zone de dépôts finiglaciaires est axée SE-NW dans le sens du ruisseau de Bury. La carte ci jointe en montre le secteur amont (en noir et blanc pour fin de reproduction).

A la tête de ce système, sur les pentes les plus faibles (0-4°), une moraine d'ablation s'est mise en place au SW et en contrebas du plus haut sommet du bassin. Au pied immédiat de ce sommet de roche en place, là où la fonte de la glace a été décélérée par la raideur de la pente (10° en moyenne), cette moraine s'intercale de plusieurs buttes et d'un bourrelet de gravier moyen.

En aval de la moraine d'ablation plusieurs générations de terrasses d'obturation sont étagées sur les longs versants à pente moyenne (4-8°) ou vers le sommet des versants à pente plus forte (plus de 8°). Les interfluves sont souvent surmontés de bourrelets ou de buttes de fusion. Les zones de pente très faible, replats ou fonds plats, sont aussi occupées par de petites buttes de fusion mais surtout par des éspandages ou des placages de sable sur moraine.

Les formes les plus importantes (bourrelets et terrasses d'obturation surtout à talus irrégulier) se sont mises en place en général à partir du moment où la pente topographique dépassait les 8°. Même si sous la plupart de ces dépôts il y a une couche de moraine de fond, celle-ci est rarement épaisse, et comme ces formes sont sises soit sur des échines, soit sur des ruptures de pente, il semble évident que ces dernières soient sculptées dans la roche en place.

Les principales formes de fusion sont donc liées soit à un éperon de roche en place (bourrelets et grosses buttes), soit à un talus de plus de 8° dans la roche en place (terrasses d'obturation). Quant aux séries de petites buttes, elles sont dues à la rapidité de la fusion à partir du morcellement d'une glace mince sur des pentes faibles.

L'écoulement des eaux au contact de la glace se faisait parallèlement aux versants et donc à la charniére des talus de terrasse d'obturation. Ainsi la plupart des terrasses sont latérales. Les directions de cet écoulement varient en raison de l'altitude de la position de la glace par rapport aux exutoires de ces eaux. Pour les dépôts à plus de 1,150 pieds d'altitude, l'écoulement se faisait de manière convergente vers le sud-ouest, dans la direction de l'exutoire du ruisseau Sherman. Lorsque le Sherman fut sur le point de cesser de fonctionner l'écoulement changea de direction et se fit vers 1'WNW d'abord, puis vers le NW dans le sens de la vallée, c'est-à-dire vers la masse de glace.

Dans la granulométrie il y a un changement marqué entre les matériaux déposés au-dessus de 1,025-1,050 pieds d'altitude et ceux déposés en dessous de cette limite. Les premiers sont en majeure partie composés de sables fins ou de gravier fin avec des poches de limons ou de gravier trés grossier parfois et souvent on retrouve des stratifications en croissant dénotant un écoulement divaguant. Les suivants sont plutôt composés de gravier moyen et grossier parfois sur une base de sable grossier et les stratifications y sont nettes. Il y aurait donc eu un écoulement général plutôt lent au-dessus de 1,025-1,050 pieds et les dépôts mis en place sont disposés à la tête du système dans une sorte d'amphithéâtre barré à la hauteur du village de Bury par deux éperons de roche en place de part et d'autre du ruisseau de Bury.

Les sommets presque plats des formes au voisinage de ces éperons suggèrent qu'il y avait là un barrage de glace au moment où l'exutoire du Sherman cessait de fonctionner. Ce barrage de glace aurait momentanément retenu un imposant volume de sédiments aussi bien sur les versants de la vallée que sur un culot de glace morte demeuré dans la dépression entre la rue McIver et le Tambs Road comme en témoigne une petite terrasse vers 960 pieds d'altitude. Le barrage s'abaissant graduellement, comme l'indique l'étagement des terrasses du versant NE, les dépôts laissés en amont ont dû être entaillés au fur et à mesure pour livrer passage aux eaux de ruissellement venant du bassin dégagé.

Les eaux ainsi évacuées et concentrées, entre la glace traînant au fond de la vallée et des versants plus rapprochés ont construit quelques terrasses de matériel plus grossier entre 850 et 900 pieds. Mais le gros des matériaux était transporté à 2 milles ou 2 milles et demi de là au contact du talus de roche en place de Bury Corner et de la dépression en aval de celui-ci.

A ce niveau les eaux ont construit des terrasses d'obturation latérales (de granulométrie décroissante) en contrebas du versant sud au contact d'une glace mince qui fondait très rapidement comme en témoigne la moraine d'ablation mise en place de l'autre côté du ruisseau entre

800 et 825 pieds sur des pentes aussi faibles que la première moraine d'ablation, soit de 0-2°.

Cette moraine d'ablation, dont le lessivage a fourni beaucoup de matériel aux dépôts qui suivent, débute une autre zone de dépôts de fusion en discontinuité avec la première.

Le fait marquant est le blocage du ruisseau de Bury par un culot de glace quasi circulaire de 3 à 4,000 pieds de diamètre. Les eaux fluviatiles venant de tout le bassin se butaient au culot et sédimentaient en amont en formant un delta à 775-800 pieds d'altitude entre le versant nord et le culot où il se termine en terrasse d'obturation. Le delta lui-même est une forme horizontale d'au moins 30 à 35 pieds d'épaisseur à stratifications de sable moyen ou de sable grossier. Ces stractifications sont inclinées à 28° vers le NW et son surmontées de 7 pieds de graviers stratifiés horizontaux en discontinuité. Au NW et au SW du culot, le delta se termine sous forme d'épandages de gravier fin ou moyen. Le ruisseau de Bury semble avoir été détourné de son cours par cet imposant volume de matériaux.

Au nord du delta, une zone de buttes de fusion et quelques petites terrasses d'obturation indiquent qu'il y avait un apport d'eau de fusion venant du nord, de la vallée du Saint-François.

En aval du delta, après un espace libre de dépôts de près d'un mille, il y a u zone de petits tertres de sable fin entre une énorme butte de fusion et un alignement de bourrelets et de buttes soulignant l'interfluve entre le ruissea de Bury et le Saint-François. Ces petits tertres de quelques pieds de hauteur sont mis en place sans ordre sur la moraine de fond. Les matériaux ont soit été déposés par l'eau sur une glace stagnante extrêmement amincie, soit par le vent avec des interstractifications de neige (Hooke, 1970), soit par une combinaison des deux processus.

Les dépôts holocènes sont peu nombreux. Ils consistent en deux cônes de déjection à la confluence du ruisseau de l'Olson's Pond et de quelques lambeaux d'anciennes terrasses en particulier à l'embouchure du ruisseau de Bury vers 675 pieds d'altitude et probablement dus à un ancien niveau du Saint-François après la déglaciation. L'Actuel est essentiellement constitué des zones alluvionnaires inondables de cours d'eau et de quelques dépôts organiques de marécages.

Il y a un fait à souligner dans la région, c'est l'existence de vallons polygéniques. Les plus gros d'entre eux, orientés NE-SW, ont été entaillés dans la roche en place au Préglaciaire par un cours d'eau à débit plus considérable que l'actuel exploitant un axe de faiblesse du substratum rocheux dans la direction de schistosité. Au Wisconsin une moraine de fond très compacte, compressée par la glace, a masqué les versants et au Finiglaciaire ils ont servi de chenal de vidange des eaux de fusion d'un bassin à l'autre (McDonald, 1969). Quant aux petits vallons, ils sont plus déterminés par la topographie que par la structure. En général, ils des gouttières parallèles aux courbes de niveau et possédant deux versants. Il arrive cependant qu'elles n'aient qu'un versant taillé à flanc de pente; celles-là étaient donc en contact direct avec la glace qui servait d'autre versant (Henderson, 1963). Ces dernières sont difficilement différenciables des simples banquettes de roche en place masquées de moraine lorsque ces dernières sont aussi parallèles aux courbes de niveau.

3 — MODALITÉS DE DÉGLACIATION:

A partir de l'étude de la nature et de l'agencement des dépôts et des formes de fusion il est possible de reconstituer les modalités de déglaciation de la dernière phase glaciaire dite de Lennoxville.

La multitude des formes de décrépitude glaciaire (Poulin, 1968), dans le bassin de Bury rend évident le fait que nous nous trouvons en présence d'une zone de stagnation. A l'échele de cette étude, nous ne retrouvons donc pas une ligne de front mais plutôt une zone de front variable selon les modalités du relief et dont l'état de récession est non seulement latitudinal mais aussi vertical selon les accidents topographiques. En effect, les endroits les plus rapidement déglacés étaient, les sommets, les barres et les saillies rocheuses représentant un amincissement de la tranche de glace et donc un pouvoir d'absorption calorifique maximum (Poulin, 1968). Ces barres ou ces interfluves en voie de déglacement délimitent des compartiments ou des bassins, préexistants dans la roche en place, faisant office d'assiettes à des culots de glace mourante vite détachés du corps principal du glacier (Poulin, 1968).

L'évolution de ces culots de glace stagnante se fait de manière individuelle et la vitesse de fusion semble être directement proportionnelle à la valeur et à l'exposition de la pente topographique (parfois la pente fait office d'écran) ainsi qu'à la charge de la glace elle-même.

Le déglacement s'étant donc fait essentiellement de manière verticale, nous pouvons donc déterminer certains niveaux de bordure de glace à partir des buttes et bourrelets de fusion et surtout à partir des terrasses d'obturation. A. Poulin (1968) a décrit ces formes et résumé cette interprétation ainsi pour la cuvette de St-Côme dans les Laurentides:

D'une part, les niveaux des terrasses indiquent l'altitude des plans d'écoulement formant les accumulations coincées entre la glace et les parois rocheuses. D'autre part, la charnière des talus indique la position limite de la glace. Les terrasses d'un même ensemble de relief, dont le niveau est approximativement de même altitude, sont donc contemporaines et la charnière des talus délimitent donc la même lentille de glace.

Son étude a cependant été faite sur une cuvette quasi fermée alors qu'ici nous avons un bassin semi-ouvert, soit un rentrant dans le versant de la vallée du Haut Saint-François. Le résultat est le même dans les deux cas excepté qu'ici le niveau de base de la zone englacée est donc externe au bassin et les plans d'écoulement des eaux de fusion ne correspondent pas aussi rigoureusement de part et d'autre la masse de glace.

4 — RÉFÉRENCES:

Bird, J. B., (1970). Some aspects of the Geomorphology of the Eastern Townships of Quebec. Rev. Géogr. Montr. XXIV, 4,pp. 417-429.

Cady, W. M., (1960). Stratigraphic and Geotectonic Relationship in Northern Vermont and Southern Quebec. Bull. Geol. Soc. Am. LXXI, 5, pp. 531-576.

Dubois, J.M.M., (1970). Bassin du ruisseau de Bury, Comte de Compton, Québec; géomorphologie et utilisation du sol, Thèse de M. A. Univ. de Sherbrooke. 137 pp.

Henderson, G. P., (1963). Etude glaciaire de la partie centrale du Québec-Labrador. C. G. C., Bull. 50, Ottawa. 96 p.

Hooke, R. Le B., (1970). Morphology of the ice-sheet Margin near Thule, Greenland. Jour. of Glaciol., IX, 57, pp. 303-324.

Klimaszewski, M., (1963). Landform list and signs used in the detailed Geomorphological Map. Inst. Geogr. Polish Acad. Sci. 56, pp. 139-179.

McDonald, B. C. , (1967). Pleistocene events and chronology in the Appalachian Region of Southeastern Quebec, Canada. Ph. D. thesis of Yale Univ. 161 pp.

McDonald, B. C., (1969). Surficial Geology of La Patrie-Sherbrooke Area, Québec. G.S.C., Ottawa, Paper 67-52, 21 pp.

Poulin, A., (1968). La Cuvette de Saint-Côme; essai de géomorphologie, Thèse de M.A., Dépt. de Géo. Univ. de Montr.

Ritchot, G., (1963). La plate-forme de Montréal; étude de cartographie géomorphologique. Thèse de 3e cycle, Univ. de Strasbourg.

COMPARISON ET INTERPRETATION DES RESULTATS DE DEUX PARCELLES EXPERIMENTALES D.EROSION 1969-70 (SHERBROOKE, QUEBEC).

Pierre Clément,
Pierre Gadbois

ABSTRACT

Two erosion experimental plots, located on a 10° slope on the campus of the University of Sherbrooke, has been under observation since the summer of 1968. Local conditions are: silty till underlying a complex soil with an Ap horizon truncating a B horizon from the former podzolised soil; grass cover following ancient cultivation. This cover has been kept on one of the plot whereas the other has been stripped bare of it.

The runoff coefficient in 1969 was 6% of the total precipitation on the bare plot and 1% on the covered one. The highest coefficients were found in May. The total sediment yield was 9,123 g. for the bare plot and almost none for the other one. The various sediment yields and the total rainfall kinetic energy are highly correlated; this gives the following regression equation = log Y = -7.35 + 1.81 log X, where Y is the sediment weight and X the total kinetic energy. Later the rainfall erosion has been strongly impeded by a partial grass colonization. Comparison of grass size distribution of eroded and soil samples from the bare plots shows a selective erosion of fine silt and clay fractions.

Durant l'été 1968, deux parcelles expérimentales d'érosion ont été installées sur un terrain mis à notre disposition sur le campus de l'Université de Sherbrooke. Les premiers résultats, obtenus en 1969-70, sont ici comparés et interprétés en fonction de l'énergie cinétique de la pluie.

LES INSTALLATIONS

Condistions du milieu:

Le versant choisi pour ces parcelles constitue le flanc sudouest d'une avancée, en partie rocheuse, formée par des schistes à séricite et des roches vertes de la formation d'Ascot. Sa pente moyenne est d'environ 10°; des ruptures de pente dues à des alignements rocheux obéissant à la structure géologique locale l'accidentent. La roche en place est recouverte par un manteau irrégulier, de 25 à 100 cm d'épaisseur, de moraine de fond limoneuse ou limonosableuse, comportant blocs et galets d'origine surtout locale. La végétation est une couverture herbacée de graminées diverses, de chardons, de verges d'or et de légumineuses; il s'agit d'une prairie semi-naturelle, due à l'abandon de terres autrefois exploitées; l'ancien climax était une forêt mixte. Le sol reflète cette évolution: les horizons de surface ont été perturbés par la mise en culture et l'horizon Ap qui en a résulté possède peu de différenciation verticale. Il repose sur l' ancien horizon B du sol précédent, qui devait être podzolique (à certains endroits, un mince horizone Ae, cendreux se retrouve). Une érosion superficielle s'est effectuée lors de la mise en culture passée, attestée par les variations granulométriques le long du versant (Gadbois, 1970).

DESCRIPTION:

Les deux parcelles expérimentales mesurent 80 m^2 de surface; elles ont la forme d'un rectangel de 15 m X 5 m, prolongé par un triangle de 2 m de hauteur. Leurs côtés sont isolés au moyen de parois de fibres de verre, n'apportant aucune perturbation chimique aux eaux ruisselées. Celles-ci sont conduites vers des cuves en acier inoxydable par des canaux en métal identique. Les volumes d'eau ruisselée ont été calculés à partir des hauteurs measurées dans les cuves. Les sédiments recueillis ont été pesés après séchage à l'air ambiant et élimination des débris organiques. L'observateur, pendant la période considérée, a laissé s'accumuler les matériaux afin de limiter la perte inhérente à leur extraction de la cuve.

Une des parcelles a été conservée sous sa couverture végétale originelle. L'autre en a été débarrassée mécaniquement; pour empêcher la repousse des herbes, des herbicides organiques (les herbicides chimiques auraient perturbé les analyses des produits dissous entraînés) ont été utilisés sur les conseils des agronomes de la Ferme Expérimentale de Lennoxville.

Les renseignements sur la pluviométrie ont été obtenus grâce au pluviographe enregistreur de la station météorologique du Département de Biologie de l'Université de Sherbrooke. Par dépouillement des bandes enregistrées, nous avons calculé les totaux et les intensités horaires des pluies. Cette information est d'autant plus valable que la dite station n'est qu'à quelques dizaines de mètres des parcelles, localisation précieuse lorsqu'on connaît la répartition capricieuse des précipitations estivales.

LES RESULTATS

Nous n'avons comptabilisé aucun résultat jusqu'au printemps 1969 afin de ne pas inclure de données affectées par les travaux d'installation. Des observations qualitatives ont montré l'absence de modifications sur la parcelle couverte; sur la parcelle découverte, des effets de délogeage et de triage par la pluie ont été notés, puis des glaçages dans les creux. A l'automne et au printemps suivant, des déplacements de particules grossières par pipkrakes et lentilles de glace se sont produits.

Ruissellement:

Le pourcentage de pluie ruisselée apparaît faible: moins de I par parcelle couverte et un peu plus de 6 pour la parcelle découverte en 1969. Les plus fortes valeurs du coefficient d'ecoulement sont au printemps par suite d'un évapo-transpiration réduite (malheureusement, les données d'evaporo-métrie n'étaient pas disponibles pour cette période); les chiffres varient entre 15 et 20% en mai. Les réserves d'eau du sol, de plus, ne sont pas encore utilisées.

En 1970, au printemps, le coefficient d'écoulement est devenu inférieur à 5% et les différences entre les deux parcelles se sont estompées. La recolonisation végétale de la parcelle découverte peut en rendre compte, mais il est possible que des fuites se soient produites à l'entrées des canaux par suite du soulèvement par le gel hivernal.

FIGURE 1

EROSION:

Durant l'année 1969, la parcelle couverte n'a libéré qu'une quantite non-mesurable de sédiments. Par contre, la parcelle découverte a perdu les quantités suivantes:

Dates de prélèvement		Poids total en grammes	Erosion en kg/ha
13	avril	111	14.0
21	mai	767	95.4
5	juin	74	8.6
18	juin	428	53.5
7	juillet	1,316	166.4
13	juillet	6,197	849.9
9	septembre	220	29.0
	Total	9.123	1,261.8

En 1970, l'érosion printanière n'a apporté que moins de 50 g, soit une érosion de moins de 6 kg à l'hectare. La recolonisation végétale de la parcelle à 80%, déjà entreprise à l'automne précédent, est la cause de cette diminution. Nous avons alors, par suite du manque d'efficacité des herbicides organiques, retourné le sol. Aucune érosion notable ne s'est produite ensuite: la porosité du sol non encore tassé et la sporadicité des pluies de l'été en rendent compte.

RELATION DE L'EROSION OBSERVEE EN 1969 ET DE L'ENERGIE CINETIQUE DES PLUIES

Nous avons utilisé pour le calcul de l'énergie cinétique de la pluie les données incluses dans l'article de Wishmeier et Smith (1958). Selon ces auteurs, l'énergie cinétique de la pluie obéit à la relation:

$$Ec = 916 + 331 \log I$$

où I est l'intensité horaire de la pluie en pouces. Les unités étant le pied/tonne par acre et par pouce de pluie, nous avons transformé les chiffres en joules par cm de pluie et pour la surface de la parcelle. Le tableau suivant donne en plus l'energie totale dépensée pendant la période précédant le prélévement des sédiments.

Date	Intensité moyenne	Energie en joules par heure pour 80 m^2	Energie totale
21 mai	2.3 mm/h	30,449	432,370
5 juin	6.2	37,980	113,942
18 juin	3.8	34,348	412,177
3 juillet	3.3	33,280	638,970
13 août	5.7	37,286	1,278,913
22 septembre	6.4	38,301	689,420

La figure 2 montre la relation entre la perte en sédiments et cette énergie totale. Nous n'avons pas utilisé pour le calcul de l'équation de régression la donnée du 22 septembre, qui se situe manifestement en dehors de la droite. A ce moment, la couverture de la parcelle par des espèces végétales pionnières freinait l'érosion. L'équation de régression est donc:

FIGURE 1

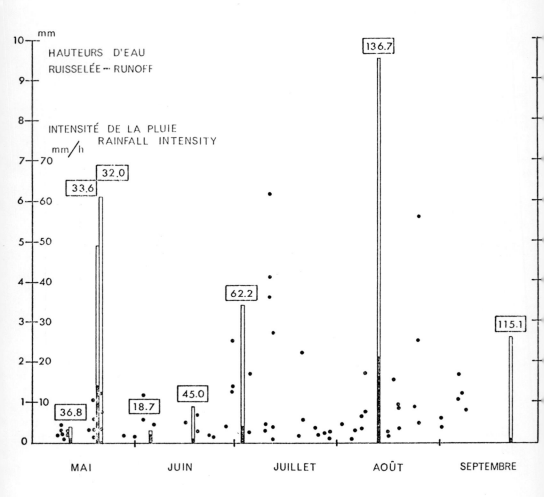

mm
10
HAUTEURS D'EAU
RUISSELÉE – RUNOFF
9
8
INTENSITÉ DE LA PLUIE
RAINFALL INTENSITY
mm/h
7—70
6—60
5—50
4—40
3—30
2—20
1—10
0

136.7
32.0
33.6
62.2
45.0
18.7
36.8
115.1

MAI   JUIN   JUILLET   AOÛT   SEPTEMBRE

PRÉCIPIPITATION TOTALE EN MM
18.7
TOTAL RAINFALL IN MM

PARCELLE COUVERTE
COVERED PLOT

PARCELLE DÉCOUVERTE
BARE PLOT

$$\log Y = -7.35 + 1.81 \log X$$

où Y est le poids de sédiments arrachés et X l'énergie totale. Le coefficient de corrélation r est de 0'988.

FIGURE 2

L'énergie cinétique totale apparaît ainsi un bon indice d'agressivité climatique, puisqu'il intègre la précipitation totale (les pluies à faible intensité préparent le sol à l'érosion en l'humectant) et les diverses intensités de celle-ci. Les pertes les plus fortes sont dues aux fortes intensités d'été, comme on le voit sur la figure 1. Des valeurs élevées au printemps sont plutôt causées par l'abondance de l'eau disponible.

L'érosion par la pluie et le ruissellement qui en résulte sont donc efficacement freinés par une couverture partielle d'herbes pionnières, à feuilles étalées sur le sol (notamment Hieracium auranticum et Taraxacum officinale) et sont annulées par un gazon continu.

LES MATERIAUX ARRACHES

L'érosion par la pluie et le ruissellement est sélective, comme l'indique la figure 3. Les deux courbes en trait continu représentent des échantillons prélevés à deux endroits différents de la parcelle découverte; le plus grossier est le plus proche du système collecteur. Le tableau suivant, établi à partir des moyennes des échantillons, montre la perte préférentielle en matériaux fins (limons et argiles) et l'enrichissement relatif résultant du sol en sables, si on ne considère que la fraction fine. La comparaison de photographies prises à des dates différentes témoigne aussi de l'accrossissement en fractions supérieures, graviers et galets. Cette observation est conforme aux résultats d'analyses des textures effectuées le long du versant voisin (Gadbois, 1970).

FIGURE 3

	Echantillons de la parcelle	Sédiments arrachés
Sables grossiers	9.8	6.9
Sables moyens	12.3	6.6
Sables fins	22.8	9.4

	Echantillons de la parcelle	Sédiments arrachés
Limons grossiers	3 4.2	3 3.7
Limons fins	13'3	28.6
Argiles	7.0	14.2

Les sédiments les plus fins ont été recueillis en juin, donc durant la période de basses intensités des pluies; les plus grossiers en avril (lors de la fonte de la neige) et lors des pluies à fortes intensités (été et automne).

CRITIQUE

Les résultats de pluviométrie couvrant la période de mai à septembre seulement, notre analyse s'est trouvée limitée. Les données concernant l'évaporation et l'humidité du sol nous ont fait défaut, les instruments n'ayant pas fonctionné convenablement ou ayant été installés ultérieurement.

D'autre part, nous n'avons pu détailler notre information par précipitation individuelle, l'ovservateur ayant laissé les sédiments s'accumuler. Cependent, les résultats nous paraissent plus fiables, puisque l'erreur du prélèvement dans les cuves a été diminuée.

Remerciements:

Nous remercisons le Comité consultatif national de la Recherche géographique pour la subvention qui a rendu cette étude possible, le Département de biologie de l'Université de Sherbrooke pour les renseignements météorologiques de sa station et le Service des bâtiments et terrains de la même Université pour son aide matérielle.

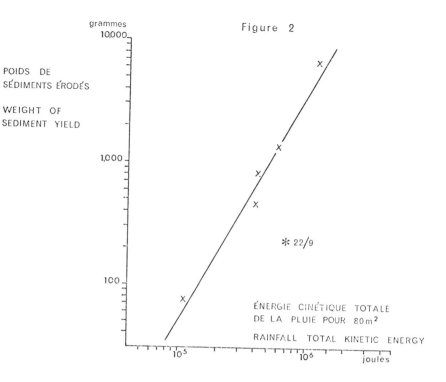

Figure 2

POIDS DE
SÉDIMENTS ÉRODÉS

WEIGHT OF
SEDIMENT YIELD

grammes
10,000

1,000

* 22/9

100

ÉNERGIE CINÉTIQUE TOTALE
DE LA PLUIE POUR 80 m²

RAINFALL TOTAL KINETIC ENERGY

10⁵   10⁶
                    joules

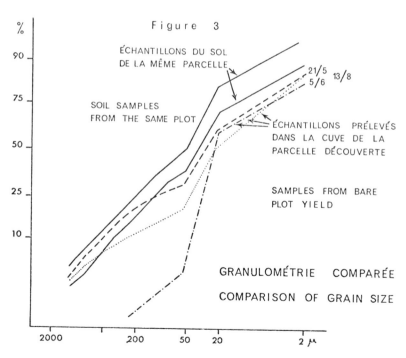

Figure 3

%

90

ÉCHANTILLONS DU SOL
DE LA MÊME PARCELLE

75

SOIL SAMPLES
FROM THE SAME PLOT

21/5 13/8
5/6

50

ÉCHANTILLONS PRÉLEVÉS
DANS LA CUVE DE LA
PARCELLE DÉCOUVERTE

25

SAMPLES FROM BARE
PLOT YIELD

10

GRANULOMÉTRIE COMPARÉE

COMPARISON OF GRAIN SIZE

2000   200   50   20   2 µ

273

REFERENCES BIBLIOGRAPHIQUES

Féodoroff, A., 1965: Mécanismes de l'érosion par la pluie, Revue de géographie physique et de géologie dynamique (2), vol. VII, fasc. 2, p. 149-163.

Fournier, F., 1960: Climat et érosion, thèse, P.U.F., Paris.

Gadbois, P., 1970: Contribution à l'étude de l'érosion d'un petit bassin versant (environs de Sherbrooke). Mémoire de Maîtrise, Université de Sherbrooke, 154p.

Wischmeier, W. H. , Smith, D.D., 1958: Rainfall Energy and its Relationship to Soil Loss, Trans, American Geophysical Union, 39, p. 285-291.

PRELIMINARY OBSERVATIONS ON DOWNSLOPE MOVEMENT OF SOIL DURING THE FALL IN THE CHINOOK BELT OF ALBERTA

Stuart A. Harris

Studies of downslope movement of material have rarely been carried out in Alberta. Campbell (1970) has shown that water erosion in the Steveville Badlands averages about three feet per thousand years. Other processes were not mentioned, and the spatial distribution of even the process of water erosion has yet to be described. In the mountains, published process studies are limited to the description of the frequency of rock falls and avalanches at specific times of the year, and also to the nature of their products (Gardner 1970a; 1970b; 1970c; Luckman, 1970). Thus there is a great need for process-oriented studies involving the spatial variation of downslope movements on a year round basis.

Accordingly, a pilot study of types of downslope movements of material at different times of the year was commenced at the Environmental Sciences Centre at Kananaskis in September, 1970. Thirteen stations were selected in a north-south traverse across the Kananaskis river valley from the shoulder of Pigeon Mountain (6000' O.D.) over to the lower spur of Barrier Mountain in the Front Range of the Rocky Mountains. This paper reports on the first results of this work, viz., the results for the Fall and Winter seasons.

Instrumentation

At each site, instrumentation includes snow stakes, rain gauges, direct-reading soil movement indicators and run-off traps. Thermocouples are used to measure soil temperature to 150 cms., while a neutron probe is used to measure soil moisture. Two Lambrecht continuous soil temperature recorders provide a check on the thermistors at 5, 10, and 20 cms. Stevenson screens containing hygrothermographs were installed at 9 key sites to check on microclimatic changes, while adjacent sites in different vegetation covers are used to monitor the effects of vegetation. Movement probes are in duplicate whenever possible and are on slopes of a variety of different angles. Snow densities are measured at each site by using a 250 cc NRC snow sampler. This network will be extended so as to provide still more information at each site.

The stations are visited once a week, and it is planned to continue this study for at least two years. Soil profiles have been described

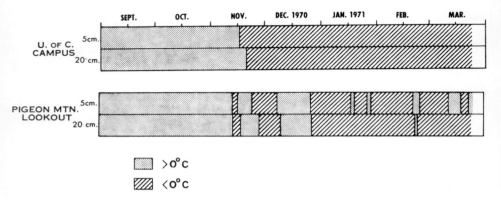

Fig. 1 Soil temperatures at 5 cms. and 20 cms. under bearberry on the shoulder of Pigeon Mountain, contrasted with those at Calgary under prairie during the same period.

at each site and samples taken for laboratory analysis. Laboratory analyses employed to date include grain size (pipette method), organic matter, free iron oxides, pH (in calcium chloride and in sodium fluoride), liquid limit, plastic limit, and X-ray diffraction of the clay-sized particles.

Environmental Factors

1. The Soils

These appear to change from regosols on the scree slopes via brown forest soils to humus podzols under deciduous or coniferous forests. Sola are thinner on north-facing slopes than on south-facing slopes (a mean of 25 cms. instead of about 40 cms.). Slopes varied from 6° to 34°. The parent materials are weathering sandstone and till which have locally been reworked by alluvial action. Clay minerals consist of calcite and dolomite from the till, and quartz, illite and chlorite from the sandstone. Textures are silt loams and loam, with sandy loams on the screes, while liquid limits ranged from 18.3 to 31.8% moisture by volume.

2. Vegetation

Vegetation cover varies from bare scree via a bearberry-rose association *(Arctostaphylos uva-ursi* (L.) Spreng. -*Rosa acicularis* Lindl.) to Aspen-aldex, *(Populus tremuloides* Michx. - *Alnus crispa* Ait. Pursh.) and thence to a Lodgepole pine-Douglas-Fir-White Spruce Forest. *(Pinus contorta* Loudon var. latifolia Engelm. - *Pseudotsuga menziesii* (Mirb.) Franco var. *glauca* - *Picea glauca* (Moench.) Voss var. *albertiana* (S. Brown) Sarg.). Past forest fires are obvious in the field since charcoal occurs in some soil profiles beneath later soils, and some charred spruce logs can be found in the open areas.

3. Climate

The area lies in the path of the intermittent Chinook winds during the long cold winter but only has about 58 frost-free days a year on average (Duffy and England, 1967). Mean annual temperature is 2.5^oC. Precipitation is about 500 mm. in the valley, of which 40% falls in the form of snow. The first snow fell in early September in 1970.

4. Microclimate

Soil temperatures (Fig. 1) were high in the fall but dropped each time it snowed. When the first heavy snowstorm came in early November, the bulk of the heat in the surface metre of soil was used in melting the snow. Since the snow cover remained thin until February the soil temperature generally follows the air temperature back and forth across $0^{\circ}C$. However the soil never dropped below $-11^{\circ}C$ and thawing was frequent. Apparently enough heat is stored at depth to keep the surface relatively warm.

The soil moisture was around 1% by volume in September. By the first heavy snowfall in November, it was averaging around 8%, but it quickly increased to about 18% as the snow melted and the water penetrated the soil (Fig. 2).

Types of Movement

So far, no runoff has been observed. Accordingly water erosion has been absent during the fall 1970 and winter 1970-71. However, it will be seen from Fig. 2 that there is probably a percolation loss which is now being investigated.

Soil creep has proven to be the major element in movement of material. Its nature will be discussed below, but the movements at a given probe may amount to 150 mm. at the soil surface in one month in the Fall. This contrasts markedly with results from other places, e.g., Ohio with a mean movement of 0.5 mm. per year (Everett, 1963).

So far, solifluction and mudflows have not been observed, although it is possible that they may occur during the spring. Heaving proved a complication as in other mountain regions but the greatest amount recorded in a week so far has been 10 mm. The ice appears to melt out at frequent intervals. Snow creep is relatively unimportant due to the low snow density (0.10 gm./cc) in winter. Where the snow is sufficiently deep, it may prove to be locally important in the Spring.

Soil Creep

Fig. 3 shows the timing of movements due to soil creep from September to March at 3 probes situated within 10 metres of one another at the same site. Environmental conditions appeared constant, yet noticeable differences in movement are discernible. The best results are obtained by averaging the results from two or three probes. The holdfasts for the indicators were anchored in bedrock.

The movements are large in September and October, including during periods when the ground remains unfrozen. Slow movement

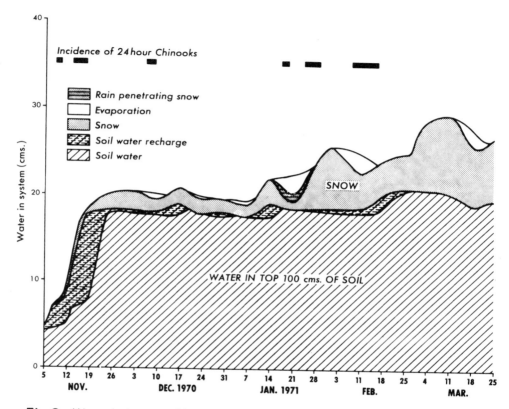

Fig. 2   Water balance at Kananaskis for November 1970 to March 1971 based on measurements at 15 sites.

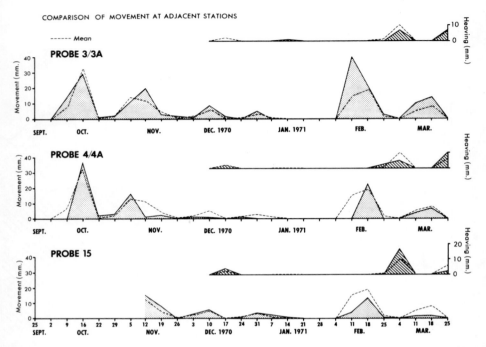

COMPARISON OF MOVEMENT AT ADJACENT STATIONS

Fig. 3 Soil creep and heaving between October, 1970 and March, 1971 on the shoulder of Pigeon Mountain at 6000' elevation under the bearberry-rose association at Site 1 as measured at three different points within 10 metres of one another.

continues until the ground is frozen solid all day. This suggests that needle ice may be involved in producing the movement at the lower temperatures. Tests are being commenced to try to establish the causes more precisely. Since the maximum movement takes place at low moisture contents (down to 3% moisture by volume) and under unfrozen conditions, flowage over a frozen layer cannot apparently by involved. Movement takes place at the same time at each site so long as environmental conditions (e.g., snow cover) are similar.

The onset of strong chinooks coupled with increasing radiation caused thawing of the soil and renewed movement at many sites periodically at the end of winter. Most sites exempt from this were sites with over 50 cms. of snow, where the latter protected the ground from the warmer conditions.

Relative movement with depth during October, 1970 was measured at six sites under different vegetation covers, using flexible tubing. There was a close correlation between vegetation cover and type of vertical profile of movement (Fig. 4). It should be noted that under forest, bedrock was too deep so that the holdfasts were driven 2 metres into the soil and the movement recorded is the differential movement between the holdfasts and the flexible tubing! Thus the figures for movement are probably minimal.

The movement is related to slope aspect and angle. North-facing slopes show less movement (a mean of 56 mm.) in the 4 months from November to February inclusive, than south-facing slopes (with a mean of 77 mm.). The movement for the winter period (5th November 1970 to 3rd March, 1971) appears to be related to the $\log^{10}$ sine of the angle of slope (Fig. 5) rather than the sine of the angle (Washburn, 1967; Schumm, 1967). The data suggests that movement may be zero on slopes of less than 2°.

Not only do the soils move, but so do some of the trees. Work is now proceeding to establish which species are involved. An aspen tree moved 66 mm. at site 2 between the beginning of November, 1970 and March 17th 1971, while the mean movement indicated by two soil movement probes within 5 metres of the tree was 72 mm. Most of the trees are shallow rooted and hence their rooting systems can survive the soil movement. Measurement of the rate of movement of trees could be a convenient way of measuring the mean soil movement rate of the rooting zone. The movement may be the reason that some tree species with marked tap roots such as the Limber pine *(Pinus flexilis* James*)* are only found on rock outcrops although they are apparently otherwise within their range

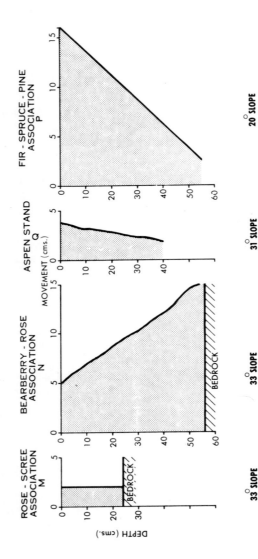

Fig. 4 Distribution of soil movement with depth from September 25th, to October 28th, 1970, under four different vegetation covers on the shoulder of Pigeon Mountain (6000' elevation).

282

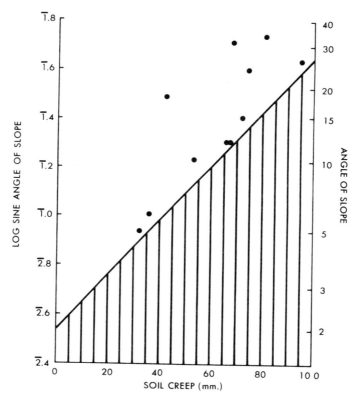

Fig. 5   Total soil creep at all stations on the south-facing slope of Pigeon Mountain between November 5th, 1970 and March 17th, 1971, as related to angle of slope.

of geographic distribution. The soil movement would tend to destroy the tap roots of this species on most slopes.

Heaving

Frost boils have been observed on scree and on bearberry covered slopes of up to 20° on the top of Pigeon Mountain. The movement indicators were arranged so that any heaving would give a reading corresponding to uphill movement. In this way, it was ensured that potential soil creep was not confused with the real thing. Heaving appears to be related to changes in soil moisture, freezing of the soil, and possibly depth of snow cover. The present data suggests that there is no simple correlation with these factors. The greatest heaving recorded on a probe to date is 25 mm., while the highest mean for all sites was 9 mm., but these low figures may be due to the unusually high snowfall this winter. Heaving appears to be far less important a process here than at higher altitudes in Colorado (Benedict, 1970).

Snow Creep

This can be shown to occur at the two sites where the snow is deepest, since the boughs of trees have been pushed downhill slightly. However elsewhere, the snow melts away from the trunks of the trees in such a way as to suggest that snow creep is ineffective in pushing over saplings. Since no tree movement was recorded when the ground was frozen, the snow could not have been moving the tree at site 2 during this period, although the snow depth reached 56 cms.

CONCLUSIONS

The main process operating on the slopes during Fall and Winter is surface creep. This movement may be the resultant of two or more processes producing similar effects under different conditions. The end-product is a steady downslope movement of the upper sheet of soil and unconsolidated material. The movement is such that on a slope of 34° under a bearberry-rose association, the movement downslope amounts to over 130 mm. between October and March inclusive. This has a variety of implications, e.g., it suggests that if unglaciated zones are present in an area with this order of soil movement the soils should not be at the surface long enough to show marked signs of age such as redness.

ACKNOWLEDGEMENTS

This study was carried out with the aid of funds from N.R.C. grant A 7483. The Alberta Forest Service loaned the author some of the necessary meteorological equipment. The writer is also indebted to B. Ciccone, C. Pilger and T. Rhodes for their assistance with the field and laboratory work.

REFERENCES

Benedict, J.B., 1970, "Downslope soil movement in a Colorado alpine region: rates, processes, and climatic significance," *Arctic & Alpine Research*, 2, 165-226.

Campbell, I.A., 1970, "Erosion rates in the Steveville Badlands, Alberta", *Canadian Geographer*, 14, 202-216.

Duffy, P.J.B., and R.E. England, 1967, *A Forest Land Classification for the Kananaskis Research Forest, Alberta, Canada*, Forest Research Laboratory, Calgary, Alberta, Internal Report A-9, 20 p.

Everett, K.R., 1963, "Slope movement, Neotormia Valley, Southern Ohio." Institute of Polar Studies, Report No. 6, Ohio State University, Columbus, Ohio, 62 p.

Gardner, J., 1970a, "Rockfall: A geomorphic process in High Mountain Terrain", *Alberta Geographer*, 6, 15-21; Gardner, J., 1970b, "Geomorphic significance of avalanches in the Lake Louise Area, Alberta, Canada," *Arctic and Alpine Research*, 2, 2; Gardner, J., 1970c, "A rate on the supply of material to debris slopes", *Canadian Geographer*, 14, 369-372.

Luckman, B.H., 1970, "The nature and variability of downslope size sorting on talus slopes," *Proc. Can. Assoc. of Geographers*, Winnipeg, pp. 213-219.

Schumm, S.A., 1967, "Rates of Surficial Rock Creep on Hill Slopes in Western Colorado," *Science*, 155, 560-562.

Washburn, A.L., 1967, "Instrumental observations of mass-wasting in the Mesters Vig District, North-east Greenland," *Medd. om Gronland*, 166, 1-297.